TOXIC STRUGGLES

The Theory and Practice of Environmental Justice

Edited by Richard Hofrichter

Foreword by Lois Gibbs

New Society Publishers
Philadelphia, PA Gabriola Island, BC

Inquiries regarding requests to reprint all or part of *Toxic Struggles: The Theory and Practice of Environmental Justice* should be addressed to New Society Publishers, 4527 Springfield Avenue, Philadelphia, PA 19143.

ISBN USA 0-86571-269-7 Hardcover
ISBN CAN 1-55092-212-2 Hardcover
ISBN USA 0-86571-270-0 Paperback
ISBN CAN 1-55092-213-0 Paperback

Cover by Parallel Design. Index by Do Mi Stauber. Book design by Martin Kelley. Printed in the United States of America on partially recycled paper using soy ink by Capital City Press of Montpelier, Vermont.

To order directly from the publisher, add $2.50 shipping and handling to the price for the first copy, 75¢ to the price for each additional copy. Send check or money order to:

in the U.S., contact:
New Society Publishers
4527 Springfield Avenue
Philadelphia, PA 19143

in Canada, contact:
New Society Publishers/New Catalyst
P.O. Box 189
Gabriola Island, BC VOR 1XO

New Society Publishers is a project of the New Society Educational Foundation, a nonprofit, tax-exempt, public foundation in the United States, and of the Catalyst Education Society, a nonprofit society in Canada. Opinions expressed in this book do not necessarily represent positions of the New Society Educational Foundation or the Catalyst Education Society.

"Building a New Vision: Feminist, Green Socialism" is adapted from the first and last chapters of Mary Mellor's *Breaking the Boundaries: Toward a Feminist, Green Socialism* (London: Virago, 1992).

"Unequal Protection: The Racial Divide in Environmental Law" is adapted from the *National Law Journal,* 21 September 1992.

"Ecofeminism and Grass-roots Environmentalism in the United States" is partly excerpted from Barbara Epstein, *Political Protest and Cultural Revolution* (Berkeley, Calif.: Univ. of California Press, 1991).

"The Effects of Occupational Injury, Illness and Disease on the Health Status of Black Americans: A Review," by Beverly Hendrix Wright and Robert D. Bullard, is adapted with permission from Westview Press from Bunyon Bryant and Paul Mohai, *Race and the Incidence of Environmental Hazards: A Time for Discourse* (Boulder, Colo.: Westview Press, 1992).

"Labor's Environmental Agenda in the New Corporate Climate" is adapted and revised from Eric Mann, "Environmentalism in the Corporate Climate," *Tikkun,* 5 (1990).

"Trading Away the Environment: Free-Trade Agreements and Environmental Degradation," is adapted and revised from an article by Mark Ritchie in *Trading Freedom: How Free Trade Affects Our Lives, Work, and Environment,* ed. John Cavanagh, John Gershman, Karen Baker, and Gretchen Helmke (San Francisco: Institute for Food and Development Policy, 1992).

GCIU

Table of Contents

Part Two
Practice and Politics: Ecological Inequities and Visions of Possibility

RACE, GENDER, AND THE ENVIRONMENT

THE HIDDEN ENVIRONMENT: CRISIS AT WORK

THE GLOBAL CONNECTION: EXPLOITATION OF DEVELOPING COUNTRIES

Acknowledgments

TOXIC STRUGGLES IS designed for students, organizers, and others interested in the burgeoning movement for environmental justice in the United States and around the world. The essayists represent a diverse group of activists, researchers, and teachers, each of whom offers a perspective that places environmental issues in a larger context of struggles for social change, locally and globally. *Toxic Struggles* is, in part, a project of the Center for Ecology and Social Justice in Washington, D.C. I dedicate the book to courageous people everywhere who must fight against ecological degradation every day as a matter of survival.

Many people provided valuable assistance and suggestions in the preparation of this book. I would like to thank Valerie Wilk, Paul Stern, Julie Barnet, Charles Frederick, Danny Faber, and Haney Khalil, each of whom patiently reviewed early drafts of some of the articles and provided insights and support, and the Lucius and Eva Eastman Fund for their financial support. I would also like to express my appreciation to Richard Healey and the Institute for Policy Studies for its generous support through in-kind services in the early phases of the project. I am deeply grateful to the late Dinny Stranahan, who supported the Center for Ecology and Social Justice with generous donations. Finally, I thank Barbara Hirshkowitz at New Society Publishers for her invaluable assistance in shepherding the book through publication as well as her patience and insightful comments on the manuscript.

Foreword

Lois Gibbs,
Citizen's Clearinghouse for Hazardous Waste

WHEN MY MOTHER asked me what I wanted to do when I grew up, I said I wanted to have six children and be a homemaker. I moved into Love Canal, and I bought the American Dream: a house, two children, a husband, and HBO. And then something happened to me, and that was Love Canal. I got involved because my son Michael had epilepsy, and my daughter Melissa developed a rare blood disease and almost died because of something someone else did. I never thought of myself as an activist or an organizer. I was a housewife, a mother, but all of a sudden it was my family, my children, and my neighbors. I believed in democracy, but then I discovered that it was government and industry that abused my rights. But my experience is not unique.

Every day across the country people distressed about the health and well-being of their families demand justice. Families who have diligently paid their taxes, farmers who have struggled for generations to keep their family farms, poor families desperately trying to break the pattern of poverty and give their children a fighting chance: they have typically suffered the greatest inequities. They are the families who make up the backbone of our country, to whom politicians play for votes, and those for whom corporations design and direct their advertising.

These are the people who make up the grass-roots movement for environmental justice. This movement, in hundreds of local and regional organizations, is typically led by women, working-class people, and people of color. Many, particularly the women, have never been involved in any political issue before and have been galvanized primarily by their concern for their children's safety. Women in Texas, for example, organized against the pollution they believed was causing brain cancer in their children. A mother of a child born severely retarded became a leader in a movement against a battery recycler in Throop, Pennsylvannia, that contaminated her neighborhood with lead. A mother of two autistic children in Lemonster, Massachusetts, discovered that there were over one hundred autistic children in her community, a statistic she believes is the result of policies at a nearby Foster-Grant sunglasses plant. The heartbreak and struggles can be heard in every state in this country.

Although these leaders became involved because of a single issue or problem, they quickly recognized the interconnections with other injustices they face daily. They realize that the root of their problem is the lack of organized political power, deteriorating neighborhood conditions, poverty, and race. Increasingly, they are beginning to recognize the international dimensions of the problem and make common cause with others around the world, such as during the United Nations Conference on Environment and Development (UNCED) in Rio de Janeiro in June, 1992. Community leaders understand, for example, that the dump or incinerator in their community is only a symptom of a larger problem: the lack of political power to resist the companies and public officials who put it there. As they battle with various bureaucracies to resolve the crisis that brought them together, they begin to identify links among issues and build an even broader coalition for change.

As a result, these leaders now build bridges with civil- rights and labor organizations, housing groups, and those fighting for adequate health care for all. As these groups join together, they develop the skills, information sources, and means of communicating the dangers they face to a wider public. These new alliances and cooperative work can achieve real democracy.

Those in power, however, fearful of these new alliances and the potential for resistance, have begun to drive wedges between the diverse groups. They are concerned about citizens rethinking how and what is produced and for whom. Such rethinking may lead to more democratic ways of decision making about ecological matters.

Community leaders are making connections with other environmental organizations and developing partnerships with local nonenvironmental groups. Civil-rights leaders, economic development groups, healthcare advocates, and women's groups are increasingly addressing more basic forms of oppression and unequal power. Alliances at the local and state level result in national coalitions. The First National People of Color Environmental Leadership Summit, held in Washington, D.C., in October, 1991, was a prime example of diverse groups with many issues devising a plan to force social change.

A major goal of the grass-roots movement for environmental justice is to rebuild the United States, community by community. Many people no longer accept the either-or rhetoric of jobs vs. the environment, health care vs. housing, or education vs. economic development. They know that there are numerous choices. This movement, working from the bottom up, builds powerful community organizations that enact strong local laws, launch ballot initiatives, and run their own candidates for office at all governmental levels.

Toxic Struggles will enrich public discussion about the developing movement for environmental justice. Reflecting a chorus of voices and perspectives, locally and internationally, it is a useful source for students, organizers, and citizens who want to know more about grass-roots struggles for environmental justice and their connections with issues of race, women's rights, economics, and occupational safety and health. Many of the contributors to this book have worked for years on issues of social justice and have first-hand experience to share with those who want to learn about these struggles and ways to change the present balance of power. I have worked personally with many of them, and I am proud to be a part of this bigger effort to win environmental and social justice for all people.

Introduction

Richard Hofrichter

AS THE ECOLOGICAL crisis deepens, a culturally diverse grass-roots movement for environmental justice grows stronger each year. This movement diverges from the essentially white, middle-class environmental movement that historically made important achievements in conservation, pollution control, species and habitat preservation, the quality of life, and consumer protection. Formerly divided by race and class, and separated by geography, dozens of grass-roots groups over the last twenty years have begun to forge new alliances to combat corporations poisoning their communities, workplaces, and children. Using strategies from the civil-rights, antiwar, antinuclear movements, as well as from other movements for social justice, these communities where people live and work are taking a leadership role in redefining the scope of the environmental movement to include social conditions that people experience in everyday life. They are making connections between undemocratic production and investment decisions, energy policies, international trade and lending policies, environmental effects of nuclear radiation and military power, and the inequities of race and class that affect the quality of their lives and the world in which they live.

For example, CATA, the Spanish acronym for the Farmworkers Support Committee in Glassboro, New Jersey—provides technical assistance and education to Puerto Rican farm workers. West Harlem Environmental Action in New York City works to limit the placement of potentially hazardous facilities and to close others that negatively affect the health of residents in the neighborhood. In rural Louisiana's "Cancer Alley," the Gulf Coast Tenants Organization fights the oil and petrochemical industries. People for Community Recovery in Chicago, the only environmental organization located in a public-housing project, opposes the legacy of abandoned toxic dumps and a plethora of industrial-waste handlers. In local communities across the country, disenfranchised poor, people of color, women, farmers, migrant farm workers, and industrial workers are making common cause with each other, and with civil-rights, peace, and women's-health groups. From the West County Toxics Coalition in San Francisco, the SouthWest Organizing Project in Albuquerque, New Mexico and Mothers of East Los Angeles to Native Americans for a Clean Environment in Tahlequah, Oklahoma, the West Dallas Coalition for Environmental Justice, and Tucsonians for a Clean Environment in Arizona, the newer grass-roots organizations are devising multi-issue, regional, and international strategies designed to overcome cultural differences and train people in community organizing, protest, and negotiation. Support from national groups such as the Commission for Racial Justice of the United Church of Christ; the Labor/Community Strategy Center in Los Angeles; the Panos Institute in Washington, D.C.; and the Citizens Clearinghouse for Hazardous Waste in Fairfax, Virginia, bolsters their organizing efforts. Multi-issue, confrontational strategies arise as these communities increasingly recognize the relationship between deteriorating conditions in their neighborhoods and the unregulated, often racist, activities of major corporations who target them for high-technology industries, incinerators, and waste.

Grass-roots groups and their allies are demonstrating to the mainstream environmental movement that issues such as protection of nature, conservation, chemical emissions, ground-water contamination, consumer product safety, free trade, exposure to pesticides, and toxic dumping are about power, self-determination, and decision making over land use, investment, working conditions, housing conditions, and energy sources. "Environmental justice demands the right to participate as equal partners at every level of decision-making including needs assessment, planning, implementation, and evaluation."[1] Their agendas go beyond regulation and incentives for business to demands for a superfund for workers, restrictions on capital flight, the elimination of the production of toxic substances, the development of a less-polluting transportation system, economic development, equitable distribution of the cost of cleanup, and international laws that protect the environment and worker rights.

A lot is at stake. The uneven distribution of resources and development that characterizes U.S. society finds a strong parallel in the distribution of ecological hazards, particularly among underrepresented, disenfranchised populations—African Americans, Latino Americans, Native Americans, Asian Americans, the poor, and women—in their workplaces and in congested urban communities.[2] A 1983 report by the U.S. General Accounting Office documented the socioeconomic and racial characteristics of communities where hazardous-waste landfills are sited.[3] Their conclusion: three-fourths are poor, African American, and Latino American. The Commission for Racial Justice of the United Church of Christ found similar patterns in its 1987 landmark study.[4] Not coincidentally, these same populations typically receive inadequate public-health and social services and live in economically underdeveloped areas with high unemployment. In a major 1991 *National Law Journal* study of civil court cases, researchers found that "[p]enalties under hazardous waste law at sites having the greatest white population were about 500 percent higher than penalties at sites with the greatest minority population."[5] The study also found discriminatory treatment in response time to hazards as well as in the effectiveness of methods in handling crises.

Poor and disenfranchised communities in the United States and in many Third-World nations are targeted for hazardous waste and chemicals from industrialized countries. Deforestation and the extraction of valuable natural resources represent a new colonialism. As transnational corporations become even more indifferent to geographic boundaries in their searches for cheap labor, compliant governments, and captive markets throughout the world, so too must the nature of political analysis and strategy shift to a global perspective that incorporates these changes.

In the United States, we have our own form of the new colonialism. People of color, concentrated in risky, low-paying jobs, find their homes located in high-pollution areas in deteriorating neighborhoods. Each year over 313,000 Latino American farm workers suffer pesticide poisoning from frequent and prolonged exposure in the fields and from the absence of field sanitation facilities.[6] Native American communities, such as the Navajo and Laguna, find their sacred lands used as sites for radioactive mill tailings, abandoned mines, or asbestos as they receive pressure from corporations to choose between using their land as dumping grounds for medical or nuclear waste, a fact that destroys their culture and sovereignty. Pollution of their water and livestock have dramatically increased their cancer rates.[7] Many African American communities, especially across the South, are disproportionately affected by ecological hazards, becoming prime sites for garbage dumps, hazardous facilities, and polluting industries. Recent studies found that race and underrepresentation on governing bodies were significant factors, independent of poverty and low income.[8]

Women, especially those whose health and survival have always been particularly affected by ecological deterioration because of gender inequalities, are rapidly becoming more politicized each year, whether as low-income residents living near a toxic-waste site or incinerator or as workers in unsafe conditions that affect their reproductive capacities. The risk of cancer for them and for their children remains high. Women in Third-World countries seek to protect natural resources necessary for survival through movements such as the rural Chipko (tree-hugging) in India. In most societies, gender influences women's social and economic position more generally, making women especially vulnerable to diseases.[9] Their experience, based on the consequences of their socioeconomic position, has led them to play an initiating and leading role in many direct action movements for a peaceful, nuclear-free society, fighting ecological destruction and calling attention to health effects through protest, research, and public education.[10] Mothers of East Los Angeles, a Latino American group that successfully fought against a toxic-waste incinerator, is a good example. Political activism for many women of color in low-income communities emerges from everyday experiences that threaten family and community, whether the issue is about lead on the playground or uranium tailings.[11]

Yet governments and multi-national corporations continue to ignore the true character of the ecological crisis. They consistently support policies driven by market forces and market incentives. More perniciously, they actively target politically weak populations as sites for toxic materials. Corporations continually threaten workers and communities into either accepting their conditions in the workplace and dumping toxics in their communities or losing their livelihoods.[12] And politicians generally refuse to address the connections between environmental deterioration and racist, sexist, and class politics.

Addressing the comprehensive character of the environmental crisis requires a radical vision and a method for evaluating the entire social system that allows us to imagine and create a different kind of society. Such a vision must incorporate the race, class, and gender dimensions of the issues. One vital component is an idea of environmental justice as defined by the seventeen principles produced by the First People of Color Environmental Leadership Summit, sponsored by the Commission for Racial Justice of the United Church of Christ, held in Washington, DC, in October, 1991 (see Appendix A). Attended by over 650 leaders from grass-roots and national organizations, participants developed agendas and strategies for responding to environmental problems as they affect people of color. A series of regional follow-up meetings were held in 1992 and 1993, the largest attended by almost two thousand people in New Orleans in December, 1992. Principles and agendas coming out of those meetings, emphasizing the totality of life conditions, form the basis of the following analysis.

ENVIRONMENTAL JUSTICE

Environmental justice is about social transformation directed toward meeting human need and enhancing the quality of life—economic equality, health care, shelter, human rights, species preservation, and democracy—using resources sustainably. A central principle of environmental justice stresses equal access to natural resources and the right to clean air and water, adequate health care, affordable shelter, and a safe workplace. The failure to satisfy such basic needs is not the result of accident, but of institutional decisions, marketing practices, discrimination, and an endless quest for economic growth. Environmental problems therefore remain inseparable from other social injustices such as poverty, racism, sexism, unemployment, urban deterioration, and the diminishing quality of life resulting from corporate activity. Urban renewal, unhealthy working conditions, highway development, investment decisions, and land-use patterns are as much environmental issues as global warming or acid rain are social issues, stemming from power over production and investment decisions.

Environmental justice concerns eliminating privilege and exploitation connected with people's health and the production and use of society's resources. Environmental injustices result, in part, from a lack of political power, and they affect the entire fabric of social life. People live near toxic-waste sites because of housing discrimination and poverty—both intimately related to the distribution of political power and resources.

Environmental issues are therefore intimately related to relations of race, class, and gender. Environmental justice requires eliminating the institutional racism that results in locating toxic wastes sites in poor, minority, or otherwise disenfranchised communities; the deteriorating conditions under which people live and work, such as farm workers' exposure to pesticides and factory workers' exposure to chemicals or radiation; and women's disadvantaged position in the work force, which leads to reproductive hazards in the workplace.

Environmental justice also means equitably distributing society's resources, creating equitable policies that support sustainable communities, and meeting basic needs for physical and psychological health. Everyone needs to breathe clean air and work in a hazard-free environment. While arguing about how to create a less-toxic production system, we cannot justify continuing to burden populations already discriminated against even more with the consequences of the existing system. Corporations must be prevented from producing toxic and hazardous substances.

Because achieving environmental justice demands major restructuring of the entire social order, a beginning point for considering basic change is a challenge to absolute property rights and the logic of industrial capitalism's emphasis on growth without limit. That is, society's productive resources and facilities cannot be permitted to be used for any purpose, regardless of the consequences, merely because of legal rights of ownership. Second is the recognition that everyone has a claim to a clean environment, not just those who can afford it. Third, the idea of security means maintaining a sustainable ecological system rather than military or economic superiority. Achieving environmental justice will require incorporating ecological issues into a larger social-justice agenda for change and considering alternative forms of economic development. Fourth is the creation of a collective bargaining process for citizens as a means to develop democratic approaches for decisions that affect everyone through the production, investment, and use of resources. That process begins with social planning that involves community residents more directly in economic development, land use, zoning, and other decisions now made with only residents' token participation. Such a process would avoid the anarchy of the capitalist marketplace and the rigidity of a centralized state bureaucracy. It could potentially serve as a countervailing power to unaccountable corporations.

Environmental justice is therefore closely related to the practice of democracy. Who will decide what to produce for whom? These decisions are about the social and economic future of the nation. Citizens need funding and resources if they are to control the destinies of their communities' land, natural resources, and technologies. Effective popular control at the local level will demand reorganizing basic institutions of commerce and culture, as well as government policies, so that citizens can evaluate proposals and determine agendas about technologies, transportation, investment decisions, and waste disposal.

For many people of color and others denied access to decision making, these issues are about survival—whether these individuals' health is destroyed by toxic wastes seeping from the ground or their cultural traditions are violated as sacred lands become sites for surface mining, logging, and development. Such issues should not be decided by the so-called private sector when the decisions have vast public and social consequences.

How could society be reorganized? What possibilities exist for replacing a society founded on the accumulation of social wealth for private purposes with one where the social wealth is held in common, for the good of communities at large? Defining needs in relation to having commodities or to a standard of living might be replaced with a definition stressing a better life for all as members of interdependent communities. Challenging market values and the globalization of capital will require questioning traditional corporate prerogatives: decisions on the right to produce and dispose of toxic waste, land use and zoning, the location of production facilities, and the fundamental principle of unlimited economic growth. This perspective moves far beyond corporate accountability to limiting corporate decision making and the right of corporations to use property as they see fit.

THREE FORCES INHIBITING THE MOVEMENT FOR ENVIRONMENTAL JUSTICE

The Media

Mass media is owned and operated by large corporate conglomerates. Populations suffering the primary consequences of environmental degeneration and the depletion of natural resources are not often in the news. Nor do the media report what these populations are doing about ecological concerns in their communities.[13] The extraordinary First National People of Color Environmental Leadership Summit in 1991 received only the most minimal coverage. Headlines abound about contaminated drinking water, leaking landfills, and violations of federal regulations. But these items are disconnected from larger issues. Investigative reporting hardly exists; we rarely hear about how money from the chemical, oil, agribusiness or electric utility industries affect research, policy, or reporting. The public receives only limited analysis of the historical, political, and social determinants of environmental conditions. Sometimes the media attribute these problems to the natural and unavoidable consequences of life in a modern industrial society. Media language and imagery present environmental crises as problems of technological failure, regulatory failure, overpopulation or individual ignorance and careless behavior. Opponents are categorized as special interests or extremists. Rarely do media reports make connections to a broad definition of environment that includes issues of civil rights, housing, employment, the quality of urban life, or the policies of global corporations. Moreover, presentations of environmental issues often divert attention from their relation to other social injustices. Corporate appropriation of green symbols, particularly in advertisements, tends to exacerbate the shallow, sporadic coverage offered by the media.

Much literature and television programming provide a great deal of advice about how individuals can protect the environment through various personal actions. Even as it assuages our consciences and offers a feeling of efficacy, emphasis on individual behavior diverts attention from political power or institutional failures.

The Mainstream Environmental Movement

The large, self-defined environmental organizations, mostly operating out of Washington, D.C., are typically referred to as the mainstream movement. It includes organizations such as the World Wildlife Fund, the Sierra Club, the Natural Resources Defense Council, the Wilderness Society, The Nature Conservancy, the Audubon Society, and the Environmental Defense Fund, which emphasize conservation, regulation, and reform. These groups rarely provide a vision or perspective beyond narrowly defined legislative, judicial, and policy alternatives. They lack comprehensive economic and political analyses that challenge the structure and undemocratic control of production and investment. Their approach has often been piecemeal, legislatively driven, and oriented to compromise at every level. Trapped in the language of cost-benefit analysis and risk management, and acting within a predetermined framework defined by government agencies, the necessary direct challenge to industry remains almost unspoken. In many cases these organizations engage in partnerships with industry, even permitting executives of polluting industries to sit on their boards. And while some advances have been made, they tend to have few connections to housing, health, and civil-rights struggles that would tie them to other major social movements. The mainstream groups rarely engage in community organizing that ties ecological problems to broader concerns. They have also run their organizations hierarchically and undemocratically, with characteristics similar to the white, male power structure of the dominant culture. The "environmental" movement, as a mostly white, middle-class entity, is not a movement for social reconstruction; it does not seek to subvert the social order to bring about the type of transformation needed beyond policy change focused on regulation or containment. Because the ecological crisis is so firmly connected to other injustices and to social disorganization, the mainstream movement is not likely to achieve its stated goals without a broader opposition to those injustices and coalitions with the newer environmental-justice movement. Nonpartisanship, consensus, and single-issue politics define core elements of the mainstream political approach. By stressing protection rather than prevention, the movement avoids the roots of the environmental crisis.

The emphasis on resource and preservation represent only part of the problem. After years of scathing criticism, the mainstream movement is beginning to respond to demands for incorporating people of color in their organizational goals and leadership positions. Whether it will share resources, promote minority issues, and otherwise follow the lead of grass-roots organizations is not yet clear. A broad-based political movement for environmental justice cannot occur without recognizing the primary role of communities of color.

The Power of Global Corporations

The continuing globalization and the expanding power of corporations—typified by polluting and dangerous technologies, the search for cheap labor and raw materials, and the unrestricted sale of goods—pose serious obstacles to environmental justice. The control over resources, the determination of the form and scale of economic organization, and corporate control over transportation systems, distribution systems, knowledge, and technology provides the global corporations with overwhelming leverage. In an effort to limit constraints against them even more, they seek total liberalization of trade policies. If fully implemented, the free-trade agreements with Mexico and Canada, which restrict action by nation-states to control their own development and environmental standards, weaken the legality of citizens' rights to protect themselves.

The federal government not only provides industry with subsidies and tax breaks but increasingly seems unwilling to regulate corporate action effectively. Communities are held hostage through either the offer of jobs or the threat to remove them unless corporations receive immunity from liability for their environmentally destructive practices. The social forces responsible for ecological destruction, embedded deep in our economic and cultural institutions, support waste, destructive competition, consumption, and the treatment of every material resource and human being as a commodity. Perhaps most important is the requirement, in a capitalist society, for endless, unlimited growth that results in the depletion of natural resources, the pollution of our habitat, and the organization of urban and rural space to suit the demands of commerce. This insatiable demand for growth is not an abusive practice that can be mitigated by reform; it is endemic to the capitalist model of development. The social costs are borne by the communities at risk and by the public. Until people challenge the fundamental principles guiding the conditions and uses of our resources and labor, ecology will remain just another policy issue.

Developing a vision of what our society can be is hindered by the nature of political life in the United States, with its issue-oriented conflicts and fragmented politics. The relationship between most public decisions and the environment, at a time when global corporations continue to centralize their power, is more crucial than ever.

A DEVELOPING MOVEMENT

Environmental justice is being achieved through a developing environmental-justice movement that links ecological and social issues in order to make strategic political interventions at vulnerable points in the system.[14] Movement leaders are creating a vision of a social order that emphasizes basic ideals of social justice regarding relations of production, priorities of allocation, and use of resources and technology. This movement connects the struggles of workers, farmers, feminists, ecologists, peace activists, and people of color to challenge corporate and state power through political, economic, and legal collective action. The environmental crisis is potentially a powerful unifying phenomenon that can link these seemingly separate constituencies.

Many organizations that fight for environmental justice at the grass roots do not see themselves primarily as environmental organizations but as social-justice organizations; their issues stress racism, sexism, and inequity. Their struggles occur on many fronts: against developers, local governments, toxic industrial processes, and the business community as a whole.

If justice is about equity and participatory democracy, then any social movement designed to achieve environmental justice must be multicultural, multi-issue, and international. This is not just a strategic question but one related to the reality of globally shared problems and the irrelevance of geographic boundaries. Free-trade agreements, for example, allow corporations to move their operations to places where labor may be more readily exploited and where environmental regulations are lax.

This book explores environmental crises as they affect particular populations by situating those crises within a social-justice framework. Environmental problems are deeply connected to basic institutions of culture and commerce, particularly the organization of production and investment and the inequities that result from that organization. If U.S. institutions are to create a nontoxic economy that does not destroy its own people in the process of production, then those institutions must be reorganized more democratically.

THE ESSAYS

The essays in this book examine a range of issues and constituencies affected by the ecological crisis around the world. Additionally, they explore the obstacles to and the possibilities for developing agendas and strategies toward a more integrated environmental-justice movement. The contributors' backgrounds vary widely, from grass-roots organizers to researchers and teachers all of whom are experts on their subjects. Some of the articles were written especially for this anthology, and some were revised and updated versions from previously published material.

The organization of the book attempts to impose order on a developing subject area with many perspectives. No settled definition of environmental justice can adequately capture its many dimensions. Therefore, the first part of the book represents a sampling of the strong links between ecological concerns and the central political and social struggles of our times. The second part, which explores struggles connected to race, gender, class, and global issues, is somewhat artificially divided, given the overlap of these areas. However, it seems a useful way to classify constituencies burdened by ecological degradation. While these categorizations reflect the organization of a growing body of literature, tensions in the movement, and an often fragmented approach to the subject, they do not typify the true interrelationships that distinguish the crisis.

The anthology's purpose is to inform readers of the varying perspectives on the relationship between ecology and social justice. With an range of voices, it presents the varied meanings and broad scope of a rapidly changing environmental-justice movement that crosses class, racial, and geographic lines. The editor and the contributors share a desire to incorporate ethnic, class, cultural, and gender concerns into the way in which the public interprets the ecological crisis. The collection is suggestive of the coming political struggles that will determine how issues are framed and resources deployed. It provides a way of looking at the ecological crisis that places these issues and resources in a social context tied to contemporary conflicts about justice, activism, and a vision of a more equitable and democratic society.

NOTES

1. *Principles of Economic Justice,* prepared for the First National People of Color Environmental Leadership Summit, Washington, D.C., October, 1991.
2. See Bunyon Bryant and Paul Mohai, eds. *Environmental Racism: Issues and Dilemmas,* A collection of papers from a University of Michigan Symposium, (Ann Arbor: University of Michigan, 1991).
3. U.S. General Accounting Office, *Siting of Hazardous Waste Landfills and Their Correlation with Racial and Economic Status of Surrounding Communities* (Washington D.C.: Government Printing Office, 1983).
4. United Church of Christ Commission for Racial Justice, *Toxic Wastes and Race in the United States: A National Report on the Racial and Socioeconomic Characteristics of Communities Surrounding Hazardous Waste Sites,* New York, 1987.
5. *National Law Journal,* "Unequal Protection: The Racial Divide in Environmental Law" (21 September 1992): S2-12.
6. Ivette Perfecto and Baldemar Velasquez, "Farm Workers: Among the Least Protected," *EPA Journal* 18 (March/April 1992): 13-14.
7. Ward Churchill, "Radioactive Colonization: A Hidden Holocaust in Native America" (unpublished paper, 1992).
8. See Bunyon Bryant and Paul Mohai, eds., *Race and the Incidence of Environmental Hazards: A Time for Discourse* (Boulder, Colo.: Westview Press, 1992); and Robert D. Bullard, *Dumping in Dixie: Race, Class, and Environmental Quality* (Boulder, Colo.: Westview Press, 1990).
9. See Vandana Shiva, *Staying Alive: Women, Ecology, and Development* (Atlantic Highlands, N.J.: Zed Books, 1989).
10. See Barbara Epstein, *Political Protest and Cultural Revolution: Nonviolent Direct Action in the 1970s and 1980s* (Berkeley, Calif.: Univ. of California Press, 1991).
11. See John Markoff, "Miscarriages Tied to Chip Factories," *New York Times,* 12 October, 1992, 1.
12. See Richard Grossman and Richard Kazis, *Fear At Work* (Philadelphia: New Society Publishers, 1982) and Eric Mann, *L.A.'s Lethal Air* (Van Nuys, Calif.: Labor/Community Strategy Center, 1992).
13. See "After Earth Day: A Survey of Environmental Reporting," *Extra!* (April/May 1992).
14. See Andrew Szasz, "In Praise of Policy Luddism: Strategic Lessons from the Hazardous Waste Wars" *Capitalism, Nature, Socialism,* 2 (February 1991); Bonnie Souza, *Justice or Ecocide: The Challenge Facing the Environmental Movement and the Opportunities for Organizing in the Pacific Northwest* (unpublished masters thesis, University of Oregon, August 1992).

Part One

The Context of Environmental Justice:

Perspectives

1

Capitalism and the Crisis of Environmentalism

Daniel Faber and James O'Connor

OVER THE LAST thirty years, the environmental movement has grown into one of the most powerful forces in U.S. politics. With the growing unity among grass-roots organizations fighting for environmental justice, linking practices and consequences of environmental degradation to deeper inequities, this movement is being powerfully transformed. But for a decade or more, the environmental movement has been under attack. Spurred by global economic restructuring and a growing economic crisis, which are rendering visible the oppositional character of U.S. capitalism to ecological protection, corporate America is leading an unparalleled assault on the achievements of the environmental movement. This article outlines the impact of this business offensive as well as possible political solutions for reversing the growing ecological crisis and crisis of environmentalism in American society.

THE HISTORY OF THE MODERN ENVIRONMENTAL MOVEMENT

Modern environmentalism in the United States partly grew out of the preservationist and conservationist movements of the late nineteenth and early twentieth centuries. These movements initially sought to protect threatened species and natural wonders from commodification and capitalization at the hands of rapacious corporations and to prevent the monopolization of natural resources by a handful of robber barons. After World War II, the traditional conservation movement became more broad-based and democratic, including actions on local green belts, residential zoning, and other public lands as well as expanded national park and national wilderness preservation systems.

These struggles by the rapidly growing and increasingly mobile urban and suburban middle classes, and also of the new urban movements of the 1960s and '70s, expanded both the quantity and quality of nature available for private and social consumption. Better wilderness and park areas, species and habitat protection, environmental quality demanded as a result of congestion, decentralized industrial siting, and pollution—these were the new middle-class issues. As such, these demands for more democratic control and regulation of environmental amenities placed the exclusionary consumerist property rights of the middle class over expansionist productionist property rights of capitalist industry, often taking the form of "no-growth" politics.

An emerging wing of the environmental movement also began organizing around issues of environmental quality and human health. Frequently led by public-health and safety organizations, these new battles sometimes linked the larger labor and environmental movements into ad hoc coalitions. Occupational health and safety became primarily a worker rather than union issue but was sometimes generalized and folded into greater demands by consumer and environmental organizations for protection against the social and ecological costs of production. By the late 1960s and early '70s, worker health and safety became a powerful political issue for many politicians in timber, coal, uranium, and other natural resource states as well as in textile-producing regions.

Although there were attempts to forge more formal organizational links between the largely middle-class preservationist-environmental amenities movement, worker health and safety organizations, and environmental-justice struggles, these branches tended to develop independently of one another in the 1960s. By the end of the decade, the preservationist-environmental amenities movement had forced passage of many new laws, including the National Environmental Policy Act (1969) and the Clean Air Act (1970). Worker militancy, combined with a lobbying coalition of more than one hundred labor, consumer, religious, and environmental organizations, managed to enact the Occupational Health and Safety Act (1970).

In this limited and often ad hoc way, middle-class and working class concerns coincided over issues of environmental quality and the exploitation of nature generally. Issues over the degradation of the environment arose around industrial, mining, and construction sites. The labor movement's fight against the exploitation of workers' health and safety intensified in the workplace and spilled over into surrounding communities. Industrial unions helped shape and pass the 1970 Clean Air Act and the 1972 Clean Water Act amendments. Environmental activists supported the 1973 strike over health and safety issues against Shell Oil by the Oil, Chemical, and Atomic Workers, as well as other coalitions with consumer organizations, labor unions, and other groups concerned with the health dangers posed by pesticides and other toxics. But despite these coalitions, the jobs vs. environment dispute would soon threaten to sabotage many of the more solid and broad-based alliances between the two movements, especially in the face of growing economic problems in the late 1970s and '80s.[1]

In the 1960s and early '70s, unionists and environmentalists also joined consumer organizations concerned with the proliferation of dangerous consumer products. Together with health and safety organizations, the environmental movement helped force an unprecedented outpouring of consumer legislation, such as the Consumer Product Safety Act of 1972. Environmentalists moved naturally into issues concerning food additives, dubious baby formulas, and unsafe automobiles while fighting to control hazardous technologies and pollution. Their efforts helped create more regulation of food processing, toy manufacturing, drugs, household chemicals, and other consumer goods, although the government stopped short of tracing harmful consumer products back to dangerous production processes and undemocratic economic and regulatory structures.

The establishment of the Environmental Protection Agency (EPA), the Occupational Health and Safety Administration (OSHA), and the Council on Environmental Quality (CEQ) was an attempt to bypass traditional federal agencies captured by corporate interests and to introduce a modicum of democracy within the state bureaucracy. By 1980, the U.S. Congress had passed twenty major laws regulating consumer products, the environment, and workplace conditions. The new federal agencies, in effect, became potential tools of the environmental movement.

By the mid-1970s, thousands of groups fighting for conservation and preservation of natural resources, local amenities, worker and community health and safety, safe energy sources, and consumer product safety formed a broad-based but very loosely organized social movement. In addition to the growth of traditional conservation organizations such as the Audubon Society, the Sierra Club, the Wilderness Society, and the National Wildlife Federation, a host of new nationally based environmental organizations also sprang forth, including the Environmental Defense Fund, the Environmental Policy Institute, Greenpeace, the Natural Resources Defense Council, Friends of the Earth, and Environmental Action, whose organizers agitated around newer and more dangerous technologies and industries. In 1975, 5.5 million people contributed financially to nineteen leading national organizations, and perhaps another 20 million to over 40,000 local groups.[2] Environmentalism had arrived as a mass-based movement.

Over the last decade, a new network of grass-roots organizations have developed to challenge the practices of ecological racism, sexism, and classism on the part of not only capital and the state, but the mainstream environmental movement as well. As seen in the First National People of Color Environmental Leadership Summit in 1991 and other organizing efforts, this growing movement for social and ecological justice seeks to address the connections between poverty, racism, sexism, and the growing crisis of environmental quality in America's urban ghettos, barrios, Native lands, and other poor rural "internal colonies," such as Appalachia. Pressing for greater economic equity and political democracy, such as the right to know about hazards facing the community, as well as for more progressive environmental policies, this movement is mobilizing women, people of color, and the poorest segments of the working class, the most likely victims of industrial pollution, toxic dumping, uranium mining, and other environmental dangers. If successful, U.S. environmentalism may finally become a truly broad-based, multiethnic movement capable of implementing a national and international strategy to challenge the abuses wrought by the capitalistic pursuit of profit and power.[3]

THE ECONOMIC CRISIS OF U.S. CAPITALISM

Since the "environmental decade" of the 1970s, the U.S. economy has experienced intensified international competition from foreign capital, especially from East Asia; during this a period, the United States has lost market after market for mass-produced consumer goods and, increasingly, for many capital goods. Furthermore, the U.S. position in this increasingly globalized capitalist economy—as a competitive producer of some raw materials, foodstuffs, and energy supplies—has been further eroded by cheaper operations overseas. In combination with weak reinvestment rates, the burden of government regulations, low productivity and profits, and the growing costs of health care, consumer product safety, and environmental protection, U.S. economic growth has been slower; only the land and credit boom of the 1980s sustained the economy until the recession of 1990.

Because of the intense competition on the world market, U.S. corporations are less able to protect their sagging profits by passing along their increased costs to consumers, one reason for the relatively low inflation rates of the 1980-90s. As a result, the first imperative of U.S. capital in this current economic crisis is *not* to increase production (since markets are not expanding) but rather to lower production costs (since markets are contracting and becoming more cutthroat). Labor costs and environmental regulations are considered by many U.S. industries to be unacceptably expensive and burdensome—expenditures that companies are seeking to reduce through massive layoffs (hence, growing unemployment), less investment in pollution control, and less concern for worker health and safety.

In the eyes of corporate America, therefore, the effects of antitoxic and other struggles on the part of the environmental, labor, and minority movements are proving increasingly toxic to short-term profitability, not to speak of long-term capital restructuring and reinvestment. However, most environmentalists often ignore, downplay, or deny the possibility that the successes of the ecology movement impair corporate economic health. A good case can be made that many environmental struggles and regulations have resulted in negative, if unintentional, effects on many sectors of the U.S. economy. In the United States, environmental regulations typically add to the costs of capital but not to revenues. Unlike new machinery that increases labor productivity and indirectly lowers the unit costs of wage goods, pollution-abatement devices and cleanup technologies usually increase costs, reduce profits, or increase prices. This is not true of all pollution-control technology. Many U.S. companies employ techniques that help use fuels and raw materials more efficiently, capture and sell waste products before they leave the factory, facilitate greater control over workers and lead to higher labor productivity, and are less damaging ecologically. However, as pointed out by self-described environmentalist Vice-President Albert Gore during the 1992 presidential election, most pollution-control devices in the United States, as compared to Germany or Japan, are added on to existing plants and equipment and fail to make industry more cost-efficient. In the copper industry, for example, the costs of compliance with the Clean Air Act, Water Pollution Control Act, solid waste disposal regulations, and worker health and safety regulations amounted to over 40 percent of total capital expenditures between 1973 and 1977, making much of the industry uncompetitive in the world market.[4]

When environmental regulation raises the costs of capital in raw materials, such as copper, and capital goods industries (or industries producing inputs for other industries), higher costs are generalized throughout the economy for most other industries. The same regulations applied to consumer goods industries affect the cost of only one or a handful of commodities. In the 1970s, nearly one-half of all capital investment was made in polluting industries (e.g., oil, petrochemicals, electrical power, strip-mining, and metalworking) supplying inputs to other industries. These sectors were hardest hit by environmental regulations, a fact which, therefore, increased the costs of capital throughout the economy as a whole. These higher costs, and many other effects on capital, have created a widespread perception in the big business community that environmentalism is increasingly raising an obstacle to corporate profitability.

THE POLITICS OF THE CORPORATE REVOLUTION AGAINST ENVIRONMENTALISM

As a result of the perception that environmentalism is a major factor contributing to the gathering economic crisis of U.S. capitalism, the Business Roundtable (an organization consisting of the chief executives of nearly two hundred of the largest and most powerful companies in the land), the National Association of Manufacturers, the U.S. Chamber of Commerce, and other political coalitions organized by large corporations have launched an unrelenting counterattack on the environmental movement since the late 1970s.[5] Heavily regulated companies, especially landed capital interests involved in real estate, finance, chemicals, oil refining, timber, and other natural resources, have poured money into antienvironmental organizations, such as the Wise Use Movement, and into election campaigns of antienvironmental candidates in both of the major parties. The aim of these efforts has been to deregulate and reprivatize the economy (i.e."liberalize" or free the market of excessive state intervention that serves the larger public interest) in the hope of restoring economic growth and corporate profits, which in political terms means the reestablishment of corporate power vis-à-vis the environmental, labor, women's, and other social movements—in effect, to transform the current economic crisis into a victory for U.S. capital.

For instance, many older regulations requiring across-the-board compliance with environmental laws are being replaced with "cost-effective" reforms—pollution taxes and credits, effluent charges, and markets for pollution rights—designed to increase capital's flexibility to meet regulation requirements but continue polluting in a profitable manner. The Clean Air Act of 1990 is one such example. Supported by then Tennessee Senator Albert Gore, a key aspect of that legislation involves the selling of pollution, which can be bought and sold on the stock market in the form of "pollution credits." Such credits allowed the Tennessee Valley Authority to exceed federal limitations on sulfur dioxide and other toxic emissions, mostly in poor minority communities in the South.

Similiarly, under policies of new federalism and the rhetoric of states' rights, responsibilities for enforcement are also being shifted from the federal to the state government. Since many states are financially unable to accommodate these commitments, the hope is that these responsibilities will be neglected or that states will engage in bidding wars with other states to attract capital to their home regions by offering more favorable investment conditions, including less environmental regulation and enforcement. One reason the current recession in states like Massachusetts is deeper than in most of the country has been because of the relocation of northern capital to the sunbelt in search of cheaper labor, lower taxes and real estate costs, and less stringent environmental regulations. One such business haven is the state of Arkansas, which has one of the highest low-wage job creation rates in the 1980-90s. As governor, Bill Clinton attracted industry to Arkansas by supporting a number of antiunion ("right-to-work") initiatives, giving numerous tax breaks to business while raising regressive taxes on the working and middle classes, and selling out on pledges to improve the workers' compensation process, to push "right-to-know" legislation, and clean up the state's environmental problems.[6] We doubt that Clinton will make a significant departure from the previous administrations in these areas.

ECONOMIC RESTRUCTURING AND THE NEW ENVIRONMENTAL CRISIS

In capitalist societies, the function of an economic crisis is to reestablish the necessary economic, social, political, and cultural conditions for renewed profitability, including new institutional arrangements congruent with new technologies, industries, and patterns of commodity demand. Most important, the goal of capital restructuring in a period of economic crisis is always to reestablish corporate discipline over social movements to increase the exploitation of both workers and nature (i.e., to extract more value from labor and nature at less cost). As a result, ecological and working conditions have worsened as organized labor and environmentalists have lost political power in the 1980s and early 1990s.

This process of capital restructuring, which includes geographical relocation, plant closures, and downsizing, facilitated in great part by the "dedemocratization" of the state, has taken a number of economic forms. First, business is externalizing more costs, spending less on the prevention of health and safety problems inside and outside of the factory, on the reduction of pollution, and on the depletion of natural resources.[7] And, as evident by growing toxic waste problems, pollution, and other social/environmental costs of capitalist production, many policy initiatives in the current crisis are actually intensifying or displacing problems they were designed to cure. For instance, higher costs arising from the EPA's regulation of the disposal of toxic wastes (based on the agency's authority under the 1976 Resource Conservation and Recovery Act) is leading some companies, exasperated by the shortage of legitimate hazardous waste hauling and disposal services, to employ organized crime to handle their wastes. Organized crime, traditionally active in garbage hauling and landfilling, is now also one of the largest toxic waste disposers in the country.[8]

Environmental regulations have failed to halt the expansion of already widespread practices of illegal hazardous waste dumping, which is cheaper than legal dumping and has more adverse health and environmental consequences. And, as in the case of clean-air regulations, which rely heavily on scrubbers attached to factory smokestacks, in the words of one government report, billions of dollars are spent "to remove pollutants from the air and water only to dispose of such pollutants on the land, and in an environmentally unsound manner."[9] This illustrates the contradictory process whereby single-issue struggles tend to displace ecological problems from one nature site to another, and in different forms, and still represent high costs to the state and to capital, as well as to people's health.

Bearing the greatest brunt of this increased pollution is the working class, particularly poor racial minorities. Since capital always seeks to pollute in ways that encounter the least political resistance, peoples and communities that have the least political power or social resources to defend themselves are most vulnerable. According to a report by the United Church of Christ's Commission for Racial Justice, three out of five African American and Latino Americans nationwide live in communities that have illegal or abandoned toxic dumps. Nationally, communities with one hazardous-waste facility have twice the percentage of minorities as those with none, while the percentage triples in communities with two or more waste sites.[10] Furthermore, a 1992 study published in the *National Law Journal* found not only that the penalties against polluters in minority and low-income areas were less severe but also that government cleanups were slower and less thorough. Native Americans are also feeling the impacts of ecological racism and classism. Some reservation lands in the Southwest are so contaminated with radioactive waste that the U.S. government has studied the possibility of declaring huge sites "national sacrifice areas," unfit for human habitation.[11]

A second form of capital restructuring involves the use by many companies of newer and typically more ecologically destructive capital goods and production processes designed to protect profit margins threatened by economic competition. Capital goods industries develop cost-reducing technologies and equipment for hard-pressed companies in both the capital and consumer goods sectors. Cost reduction is foremost in the minds of business leaders who order this technology and equipment (e.g., manufacturers of new synthetics and plastics, new genetic strains, and more powerful strains of pesticides and chemicals, which have caused new problems for people and nature alike).[12] Thanks to the demand for cost-reducing technology, the disaster of high-tech pollution in places like the Silicon Valley, Boston Harbor, and the greater Houston metropolis are partly attributable to the growing crisis of U.S. capitalism and the thirst for profits. Minorities, workers and the middle class bear the greatest harm from these technologies and also pay for the cleanup.

Third, "reforms" in the Food and Drug Administration, Consumer Product Safety Commission, and offices in the Department of Agriculture, provoked by the economic crisis, allow the introduction of more profitable but also more dangerous foods, drugs, appliances, and other consumer goods, which cause an estimated 28,000 deaths and 130,000 serious injuries each year.[13]

Fourth, coupled with the government's offensive against conservationist tendencies within natural resource agencies, there are now cheaper sources of renewable and nonrenewable resources for American capital, such as the majestic old-growth forests of the Pacific Northwest, habitat for the endangered spotted owl; the rich deposits of low-sulfur coal that lie underneath the Black Mesa homelands of the Hopi and Navajo Indians in the Four Corners region of the American Southwest; or the vast oil reserves that lie in the Arctic National Wildlife Refuge in Alaska, as well as along the southern and western coastlines of the continental United States. Schemes to open new oil and coal fields and public forests are motivated less by oil, coal, or timber shortages than by energy and timber companies' need to bring in lower cost oil, coal, timber and other fuels and raw materials. The result has been pressure to expand offshore drilling, strip-mining, and destructive timber harvests with all their attendant adverse environmental consequences.[14]

Fifth, the growth of home work and new industrial sweat shops, where health and safety conditions are more or less unregulated, are also accelerating. An example is the chicken processing industry, one of the fastest-growing businesses in a number of southern states, including Arkansas and North Carolina. On September 3, 1991, a fire ripped through the Imperial Food Products poultry-processing plant in Hamlet, North Carolina, killing twenty-five workers and injuring another fifty-five. According to the Government Accountability Project, because the plant had never been inspected in eleven years of operation, safety violations were rampant (e.g.,emergency doors were locked from the outside to prevent theft, thus trapping workers inside during the fire).[15] Furthermore, some 100,000 deaths occur nationwide each year as a result of exposure to toxic chemicals in the workplace, many of those deaths of nonunion female workers.[16]

Finally, the legislative battles waged by labor and environmentalists in the United States brought about regulations in the 1960-70s, which indirectly contributed to the internationalization of capitalist production and the consequent export of environmental degradation and health problems to the Third World and elsewhere. These include the export of more profitable yet more dangerous production processes, consumer goods, and waste disposal, as well as the widespread destruction of the world's forests, and new threats to biodiversity and to the integrity of the globe's climate, oceans, and atmosphere.[17]

THE CRISIS OF ENVIRONMENTALISM

By not designing a comprehensive political and economic strategy to combat processes of capital restructuring and to develop radical alternatives to capitalism, the environmental and labor movements have failed to halt the growing environmental and health crises of the 1980-90s. The failure to push hard for ecological and labor internationalism in opposition to capitalist exploitation has proven especially disastrous in the Third World, where labor and environmental protection is weak. Ecological racism causes damage to African American and other oppressed minority communities at home, and even more so under conditions of cost-cutting and economic crisis. Environmentalism's single-issue, legislative approach has led capital to displace costs in different forms from one site and country to others. The movement's weak analysis of how and why capitalism works has helped lead to unintended, adverse effects on the well-being of people and their environments. Also, regional and local movements and coalitions by and large have not looked beyond their own areas to analyze the effects elsewhere of their own local or regional successes in environmental protection. Many of the environmental movement's legislative victories of the 1960-70s thus have become one source of its failures in the 1980-90s.

With some notable exceptions, environmental quality in the United States is worse today than it was in the 1960s. Since 1982, air pollution such as dust particles, sulfur dioxides in the form acid rain, and carbon monoxide evidenced by respiratory disease have all increased.[18] There are currently some 30,000 to 50,000 toxic-waste disposal sites, of which only 200 are licensed, with thousands more being discovered each year.[19] While more than $100 billion has been spent as a result of Clean Water Act regulations, most U.S. rivers suffer from poor water quality. More water tables are depleted or poisoned; more soil is eroded or rendered unproductive because of salinization and other problems. The occurrence of nitrate, arsenic, and cadmium, all serious pollutants, has increased considerably.[20] Ocean and shoreline pollution has become a national scandal, as has the threat to wetlands and wildlife sanctuaries. While the environmental movement has consolidated some important victories (e.g., some habitat restoration, much wilderness preservation, a ban on leaded gasoline), the movement has failed to halt the overall process of environmental destruction in the

United States. The result is that environmental, labor, minority, and community groups find themselves fighting new forms of environmental degradation as well as the reappearance of older destructive practices. Meanwhile, the economic crisis also poses a second contradiction for capital itself: The destruction of nature as a short-term crisis resolution strategy threatens to create much greater economic problems for capital in the long run.[21]

The focus on technical-rational questions, solutions, and compromises, rather than on issues of political power and democratic decision making, has recently caused a growing split between grass-roots and professional organizations. This split, apparent since the mid-1980s, has largely been inspired by the issue of professionalization, which developed not only because the general emphasis shifted from law making to law implementation but also because of the new hostility to ecological politics in general on the part of business and the federal government. Professionalization is largely a liberal strategy to reach compromises with, and win concessions from, an unfriendly national government.

The reaction at the base consists of a revitalized grass-roots politics by local and national organizations such as Earth First!, Citizen's Clearinghouse for Hazardous Waste, the SouthWest Organizing Project, Concerned Citizens of South Central Los Angeles, and People for Community Recovery in Chicago—and includes the use of direct action against timber companies, polluters, and others as well as criticism toward much of the mainstream's environmental politics, especially its ecological racism and classism. This includes the emergence of a new movement for social and environmental justice led by African Americans and other oppressed minorities. The environmental movement's internal conflicts in the 1980s and '90s are between new direct-action groups, which want to dramatize examples of environmental destruction and stop them at their source, and the politics of institutional consensus, compromise, and professionalization. The grass-roots organizations offer the potential for democratizing and transforming the environmental movement into a truly mass-based movement, inclusive of all races, the working poor, and women, that combines issues of social and ecological justice. However, those organizations and labor coalitions that ignore the connections between the political crisis of interest-group politics and the combined and interrelated force of economic and ecological crises feed divisions within the movement. At stake is not only environmental, worker, and community health, but also the viability of the traditional political strategies, tactics, and vehicles used by environmentalists and other social movements.

TOWARDS A LEFT ECOLOGY MOVEMENT

The environmental movement as a whole and the labor movement have inadvertantly been weakened by failing to revive the struggle to democratize the state and the workplace, to fight against ecological racism and incorporate oppressed racial minorities and broader segments of working people, and to develop environmental solidarity with movements in the Third World that know that capitalist economic development, ecological degradation, and human poverty are different aspects of the same problem. The growing ability of multinational corporations and financial institutions to evade and dismantle unions, environmental safeguards, and worker-community health and safety regulations in the United States is being achieved by crossing national boundaries into politically repressive and economically oppressive Third-World countries.[22] As a result, peoples and governments of the world are increasingly being pitted against one another in a bid to attract capital investment, leading to one successful assault after another on labor and environmental regulations seen as damaging to profits. In this context, using the rhetoric of free trade, free enterprise zones, and jobs vs. the environment, capital has further weakened and divided the United States' social movements against one another.

As the crisis of capitalism deepens, and global ecological conditions worsen, the need for a mass-based international Left ecology movement that unites struggles for economic, social and ecological justice will become more pressing. Just as in the 1930s, when the labor movement was forced to change from craft to industrial unionism, so today labor needs to transform itself from industrial unionism into an international, conglomerate union (inclusive of women and all racial-ethnic minorities) merely to keep pace with the restructuring of international capital. And just as in the 1960s, when the environmental movement changed from a narrowly-based conservation-preservation movement to include the middle class and sections of the working class, so today it needs to change from single-issue local and national struggles to a broad-based international movement. Unless the environmental and other social movements in the United States realize that strong environmentalism and worker health and safety concerns are needed throughout the rest of the world, there will be no protection of local initiatives and gains. Fortunately, as seen in the Global Forum at the 1992 United Nations Conference on Environment and Development (UNCED) in Rio de Janeiro, there are signs that a progressive international ecology movement is developing that combines concerns for both social-economic and ecological justice. Furthermore, there is a growing recognition that only through a democratization of both the economy *and* the state can comprehensive solutions to the social and environmental crisis of U.S. capitalism be found. This is the major challenge facing left environmental activists today.

NOTES

1. For a historical account of these convergences and capitalist-created division between the labor and environmental movements, see Richard Kazis and Richard Grossman, *Fear at Work: Job Blackmail, Labor and the Environment* (New York: Pilgrim Press, 1982).
2. See Francis Sandbach, *Environment, Ideology, and Policy* (Montclair, N.J.: Allanheld, Osmun, 1980), 13.

3. See Robert Bullard ed., *Confronting Environmental Racism: Voices From the Grassroots.* (Boston: South End Press, 1993). See also Mark Dowie, "American Environmentalism: A Movement Courting Irrelevance," *World Policy Journal*, 9 (Winter 1991-92): 67-92.
4. See Chibuzo Nwoke, *Third World Minerals and Global Pricing: A New Theory* (Atlantic Highlands, N.J.: Zed Press, 1987), 175-81.
5. See Val Burris, "The Political Partisanship of American Business: A Study of Corporate Political Action Committees," *American Sociological Review*, 52 (December 1987): 736-37.
6. For a discussion, see Alexander Cockburn, "Clinton, Labor and Free Trade: The Executioner's Song," *The Nation*, 2 November 1992, 489-509. The records reveal that, under Clinton's tenure, the Arkansas Pollution Control and Ecology Department botched inspections, failed to enforce federal standards, and had one of the worst records in the country. See Alan Levin and Nick Tate in the *Boston Herald*, 22 October 1992, 1-5.
7. In the United States, capital spending on domestic pollution-control equipment declined from 5.6 percent in 1976 to 2.7 percent in 1985. As a result, since 1982 there has been an average annual increase of .87 percent in air-pollutant emissions. See Barry Commoner, "A Reporter at Large: The Environment," *The New Yorker*, 15 June, 1987, 54; and Larry Everest, *Behind the Poison Cloud: Union Carbides's Bhopal Massacre* (Chicago: Banner Press, 1985), 36.
8. See Alan A. Block and Frank R. Scarpitti, *Poisoning for Profit: The Mafia and Toxic Waste in America* (New York: William Morrow, 1982).
9. See Lewis Regenstein, *How to Survive in America the Poisoned* (Washington, D.C.: Acropolis Books, 1986), 160.
10. See United Church of Christ Commission for Racial Justice, *Toxic Wastes and Race in the United States: A National Study of the Racial and Socioeconomic Characteristics of Communities with Hazardous Waste Sites* (New York: United Church of Christ, 1987); Robert D. Bullard, *Dumping in Dixie: Race, Class and Environmental Quality* (Boulder, Colo.: Westview Press, 1990); and Devon Pena," "The 'Brown' and the 'Green:' Chicanos and Environmental Politics in the Upper Rio Grande," *Capitalism, Nature, Socialism*, 3 (March 1992): 79-103.
11. See Ward Churchill and Winona LaDuke, "Radioactive Colonization and the Native Americans," *Socialist Review*, 81 (1985).
12. See Jeremy Rifkin, *Declaration of a Heretic* (Boston: Routledge and Kegan Paul, 1986).
13. See Joan Claybrook and the staff of Public Citizen, *Retreat from Safety: Reagan's Attack on America's Health.* (New York: Pantheon, 1984), 58-60.
14. See Patrick Mazza, "The Spotted Owl as Scapegoat," *Capitalism, Nature, Socialism*, 4 (June 1990): 98-107; and Kathy Hall, "Changing Woman, Tukunavi and Coal: Impacts of the Energy Industry on the Navajo and Hopi Reservations," *Capitalism, Nature, Socialism*, 3 (March 1992): 49-78.
15. See the Community Environmental Health Program, *Environment and Development in the USA: A Grassroots Report for UNCED* (New Market, Tenn.: Highlander Research and Education Center, 1992).
16. See John O'Connor, "The Toxics Crisis," in Gary Cohen and John O'Connor eds., *Fighting Toxics* (Washington, D.C.: Island Press, 1990), 13; and Susan Klitzman, Barbara Silverstein, Laura Punnett, and Amy Mock, "A Women's Occupational Health Agenda for the 1990's," *New Solutions*, 1, (Spring 1990): 7-17.

17. See Barry Castleman and Vicente Navarro, "International Mobility of Hazardous Products, Industries, and Wastes," *Annual Review of Public Health,* 8 (1987): 1-19.
18. See Barry Commoner, "A Reporter at Large: The Environment," *The New Yorker,* 15 June 1987, 46-71.
19. See Regenstein, *How to Survive,* 137.
20. See Commoner, "Reporter at Large," 51.
21. See James O'Connor, "Capitalism, Nature, Socialism: A Theoretical Introduction," *Capitalism, Nature, Socialism,* (Fall 1988): 11-38; and "Conference Papers on Capitalist Nature," pamphlet 1 (Santa Cruz, Calif.: Center for Ecological Socialism, 1991).
22. See Daniel Faber, *Environment Under Fire: Imperialism and the Ecological Crisis in Central America* (New York: Monthly Review Press, 1992).

2

Anatomy of Environmental Racism

Robert D. Bullard

DESPITE THE MANY federal laws, mandates, and directives by the federal government to eliminate discrimination in housing, education, and employment, government rarely addresses discriminatory environmental practices. People of color (African Americans, Latino Americans, Asian Americans, and Native Americans) are disproportionately affected by industrial toxins, dirty air and drinking water, and the location of noxious facilities such as municipal landfills, incinerators, and hazardous-waste treatment, storage, and disposal facilities.[1]

IMPACT OF ENVIRONMENTAL RACISM

All communities are not created equal. Some are subjected to all kinds of environmental assaults. Many differences in environmental quality between communities of color and white communities result from institutional racism. Institutional racism influences local land use, enforcement of environmental regulations, industrial facility siting, economic vulnerability, and where people of color live, work, and play. Environmental racism is just as real as the racism that exists in housing, employment, and education.

The roots of institutional racism are deep and difficult to eliminate.[2] Discrimination is a manifestation of institutional racism. Even today, racism permeates nearly every social institution. Environmental institutions of both governmental and nongovernmental bodies are no exceptions. Racism influences the likelihood of exposure to environmental and health risks as well as accessibility to health care.

People-of-color communities have borne a disproportionate burden of this nation's air, water, and waste problems as well as the siting of sewer treatment plants; municipal landfills; incinerators; hazardous-waste treatment, storage, and disposal facilities; and other noxious plants. Residents of many of these same communities live in housing contaminated with lead, whose problems are further complicated by hospital closures and inaccessible health clinics.

People of color have been systematically excluded from (or allowed minimal participation in) decision-making boards, commissions, and staffs of governmental bodies. Business elites promote jobs, an expanded tax base, and economic development as selling points for local residents to accept risky industries. Jobs are often promoted over the environment, especially when the community discovers how many jobs are created, the skills required, the pay scale, and *who* will actually end up getting the jobs.

Environmental racism disadvantages people of color while providing advantages (i.e., privileges) for whites. A form of illegal exaction forces people of color to pay costs of environmental benefits for the public at large. Determining who pays and who benefits from our current urban and industrial policies is central to an analysis of environmental racism. Exclusionary zoning and unequal protection have created environmental sacrifice zones where residents pay with their health. Racial barriers in housing limit mobility options available to people of color.

Racism influences every social and economic strata of people of color. Moreover, environmental inequities do not result solely from differences in social class. In the United States, race interpenetrates class and creates special health and environmental vulnerabilities. People of color are exposed to greater environmental hazards in their neighborhoods and on the job than are their white counterparts. Studies find elevated exposure levels by race, even when social class is held constant.[3] For example, research indicates race to be independent of class in the distribution of air pollution,[4] contaminated fish consumption,[5] location of municipal landfill and incinerators,[6] abandoned toxic-waste dumps,[7] and lead poisoning in children.[8]

Lead poisoning is a classic example of an environmental health problem that disproportionately affects children of color at every class level. Lead affects between three and four million children in the United States—most of whom are African American and Latino Americans who live in urban areas. Among children five years old and younger, the percentage of African American children who have excessive levels of lead in their blood far exceeds the percentage of whites who do at all income levels.[9]

The federal Agency for Toxic Substances Disease Registry (ATSDR) found that, for families earning less than $6,000, 68 percent of African American children had lead poisoning, compared with 36 percent for white children. In families with income exceeding $15,000, more than 38 percent of African American children suffer from lead poisoning, compared with 12 percent of whites.[10] Even when income is held constant, African American children are two to three times more likely than their white counterparts to suffer from lead poisoning.

People of color do not have the same opportunities as whites to escape unhealthy physical environments.[11] Most environmental-justice activists challenge an environmental ethic that allows individuals, workers, and communities to accept health risks others can avoid by virtue of their skin color. For example, African Americans, no matter what their educational or occupational achievement or income level, experience greater environmental threats because of their race.[12]

Institutional barriers such as housing discrimination, redlining by banks, and residential segregation prevent African Americans from buying their way out of health-threatening physical environments. The ability of an individual to escape a health-threatening physical environment usually correlates with income. However, racial barriers complicate this process for millions of African Americans.[13] An African American who has an income of $50,000 is as residentially segregated as an African American on welfare.

Some communities, located on the "wrong side of the tracks," receive different treatment in the delivery of public services, including environmental protection. In the heavily populated South Coast air basin of Los Angeles, for example, over 71 percent of African Americans and 50 percent of Latino Americans reside in areas with the most polluted air, while only 34 percent of whites live in highly polluted areas.[14]

THE DUMPING GROUNDS

Apartheid-type housing and development policies limit mobility, reduce neighborhood options, diminish job opportunities, and decrease environmental choices for millions of Americans.[15] Why do some communities get dumped on and others do not? Waste generation directly correlates with per-capita income. Therefore, public officials neither propose or locate many waste facilities in the suburbs.

The Commission for Racial Justice's landmark study, *Toxic Wastes and Race,* found race to be the most important factor (i.e., more important than income, home ownership rate, and property values) in the location of abandoned toxic-waste sites.[16] The study also found that three out of five African Americans live in communities with abandoned toxic-waste sites; 60 percent (fifteen million) African Americans live in communities with one or more abandoned toxic-waste sites; three of the five largest commercial hazardous waste landfills are located in predominantly African American or Latino American communities, accounting for 40 percent of the nation's total estimated landfill capacity; and African Americans are heavily overrepresented in the population of cities with the largest number of abandoned toxic-waste sites, a list that includes Memphis, St. Louis, Houston, Cleveland, Chicago, and Atlanta.[17]

In addition to racial composition, economic factors combine to increase the likelihood of a community hosting a hazardous-waste incinerator. A 1990 Greenpeace report, *Playing with Fire,* found that the minority portion of the population in communities with existing incinerators is 89 percent higher than the national average; communities where incinerators are proposed have minority population 60 percent higher than the national average; average income in communities with existing incinerators is 15 percent less than the national average; property values in communities close to incinerators are 38 percent lower than the national average; and in communities where incinerators are proposed, average property values are 35 percent lower.[18]

Garbage dumps are not randomly scattered across the landscape. These facilities are often located in communities that have high percentages of poor, elderly, young, and minority residents.[19] In 1979, one of the first studies to link race with the location of municipal solid waste sites focused on Houston.[20] From the early 1920s to the late 1970s, all of the city-owned municipal landfills, and six out of eight of the garbage incinerators, were located in African-American neighborhoods.

From 1970 to 1978, three out of four of the privately owned landfills that were used to dispose of Houston's garbage were located in African American neighborhoods. Although African Americans made up only 28 percent of Houston's population, 82 percent of the municipal landfill sites, public and private, were located in African American neighborhoods.

Siting inequity is not confined to Houston. African American communities from South Central Los Angeles to the southeast side of Chicago to West Harlem are vulnerable to waste facility siting. As recently as 1991, Residents Involved in Saving the Environment (RISE), a biracial community group challenged the King and Queen County, Virginia, Board of Supervisors for selecting a 420-acre site for a regional landfill, located in a primarily African American community. King and Queen County population is nearly evenly split between African Americans and whites. The group charged the board with racial discrimination in landfill siting and zoning since all three of the county-run landfills are located in predominantly African American communities.

In June, 1991, a U.S. district judge for the Eastern District of Virginia in *RISE v. Kay* ruled that the selection of the mostly African American area in King and Queen County did not violate the equal-protection clause, despite the county's historical placement of landfills in African American areas.[21] Although the court acknowledged that the placement of landfills in the county from 1969 to 1991 had a disproportionate impact on African American residents in the county, it failed to find discrimination.

Siting inequities are not unique only to facilities for dumping household garbage. The southern United States, our own Third World, is rapidly becoming the dumping ground for household garbage and hazardous waste. Historically, the South scored at or near the bottom on almost all indicators of well-being (e.g., education, income, economic development, environmental quality, and health care). The region has a long history of exploitation of land and people, especially African Americans, dating from slavery. There is a clear link between the region's lax enforcement of regulations designed to protect public health and the environment, lax enforcement of laws designed to protect the civil rights of its African American citizens and race relations.

Findings in *Dumping in Dixie* show that African Americans in the South bear a disparate burden in the siting of hazardous-waste landfills and incinerators, lead smelters, petrochemical plants, and a host of other noxious facilities.[22] South Louisiana's "Cancer Alley" and Alabama's "blackbelt" epitomize disparate waste facility siting.

Emelle, Alabama, hosts the nation's largest commercial hazardous-waste landfill, dubbed the "Cadillac of dumps." In Emelle, a rural community in the heart of Alabama's "blackbelt," African Americans make up over 90 percent of the population and 75 percent of the residents in Sumter County. The Emelle landfill receives wastes from Superfund sites and wastes from all forty-eight contiguous states.

Dallas, on the other hand, has a long history of allowing lead smelters to be sited in African American and Latino American neighborhoods. The dangers of lead have been known since the Roman era. The lead contamination problem in the mostly African American West Dallas neighborhood was documented by the Dallas Health Department as far back as 1969. A 1983 federal study established that the local smelter was the source of elevated blood lead levels in children who lived in West Dallas.[23] Cleanup delays by the EPA has amounted to "waiting for a body count."[24] One wonders if the residents of the mostly white North Dallas neighborhoods would receive the same treatment as the residents of West Dallas.

Comprehensive cleanup activity began in the West Dallas site in January, 1992—nearly twenty years after the first published report of the problem. An estimated 30,000 to 40,000 cubic yards of lead-contaminated soil will be removed from several West Dallas sites, including school property and the yards of some private homes. The soil is scheduled to be dumped at the Magnolia landfill in Monroe, Louisiana, a community that is over 60 percent African American.[25]

Siting inequities were identified by the U.S. General Accounting Office (GAO) nearly a decade ago. After protests sparked by the siting of a PCB landfill in the mostly African American Warren County, North Carolina, the GAO initiated its own investigation of a hazardous waste facility siting in the EPA's Region IV. The government agency found a strong relationship between the location of off-site hazardous-waste landfills and race and socioeconomic status of the surrounding communities in the EPA's Region IV.[26]

The GAO identified four off-site hazardous-waste landfills in the eight states (Alabama, Florida, Georgia, Kentucky, Mississippi, North Carolina, South Carolina, and Tennessee) that comprise the EPA's Region IV. The four sites included Chemical Waste Management (Sumter County, Alabama), SCA Services (Sumter County, South Carolina), Industrial Chemical Company (Chester County, South Carolina), and the Warren County PCB landfill (Warren County, North Carolina). African Americans made up the majority of the population in three of the four communities with off-site hazardous-waste landfills.

In 1983, African Americans were clearly overrepresented in communities with waste sites since they made up only about one-fifth of the region's population, and three-fourths of the landfills were located in African American communities. Siting imbalances that were present in 1983 have not disappeared. In 1992, African Americans still make up about one-fifth of the population in Region IV. However, the two currently operating off-site hazardous-waste landfills in the region are located in ZIP codes where African Americans are a majority of the population.

CASE STUDIES FROM CALIFORNIA

African American communities in the southern United States are not the only communities of color experiencing environmental racism. In Los Angeles, for example, the mostly African American South Central Los Angeles and Latino American East Los Angeles neighborhoods were targets for municipal solid-waste and hazardous-waste incinerators, respectively.

Los Angeles, the nation's second largest city (with a population of 3.5 million persons) is one of the most culturally and ethnically diverse cities in the United States. People of color—Latino Americans, Asian Americans, Pacific Islanders, African Americans, and Native Americans—now constitute 63 percent of the city's population. Eight of every ten of the city's African Americans and about half of Los Angeles's Latino Americans live in segregated neighborhoods. The now riot-torn South Central Los Angeles is one of these segregated enclaves that has suffered from years of systematic neglect, infrastructure decay, high unemployment, poverty, and heavy industrial use.

The South Central Los Angeles neighborhoods suffer from a double whammy of poverty and pollution. A recent article in the *San Francisco Examiner* described the ZIP code in which South Central Los Angeles lies (90058) as the "dirtiest" in the state.[27] The 1990 population in the ZIP code is 59 percent African American and 38 percent Latino American. Abandoned toxic-waste sites, freeways, smokestacks, and waste water pipes from polluting industries saturate the one-square-mile area. The neighborhood is a haven for nonresidential activities. More than eighteen industrial firms in 1989 discharged more than 33 million pounds of waste chemicals in this ZIP code.

Why has South Central Los Angeles become the dumping ground of the city? Local government decisions are in part responsible. Trying to solve them, the city (under a contract with the EPA) developed a plan to build three waste-to-energy incinerators.[28] Odgen-Martin was selected to build the incinerators dubbed LANCER (Los Angeles Energy Recovery). The first of the three incinerators, LANCER 1, was slated to be built in South Central Los Angeles.

Proponents of LANCER 1 attempted to speed the project through and locate it in South Central, as one way to blunt public opposition when LANCER 2 and 3, planned for the wealthier and mostly white Westside and San Fernando Valley came up for review. City officials reasoned that they would be hard-pressed to justify killing LANCER 2 and 3, if LANCER 1 was up and running.[29] City council members, however, underestimated the organizing skills of South Central Los Angeles residents.

After learning about the incinerator project in 1984, residents organized themselves in a group called Concerned Citizens of South Central Los Angeles, most of whom were African American women. Local activists from Concerned Citizens were able to form alliances with several national and grass-roots environmental groups, as well as with public-interest law groups to block the construction of the city-initiated municipal solid-waste incinerator. Concerned Citizens was assisted by Greenpeace, the Citizen's Clearinghouse for Hazardous Waste, the Center for Law in the Public Interest, the National Health Law Program, the Institute for Local Self-Reliance, the California Alliance in Defense of Residential Environments (CARE), and a group called Not Yet New York. Opponents applied pressure on city officials, including Mayor Tom Bradley. In 1987, the mayor and the Los Angeles city council killed the LANCER project—a project that had included a commitment of $12 million.[30]

Just as Los Angeles's largest African American community was selected for the city's first state-of-the-art municipal solid-waste incinerator, the state's first state-of-the-art hazardous-waste incinerator was slated to be built near East Los Angeles, the city's largest Latino American community. Officials of the California Thermal Treatment Services (CTTS) planned the hazardous-waste incinerator for Vernon, an industrial suburb that has only 96 people. Estimates indicated that the incinerator would burn about 22,500 tons of hazardous waste per year.[31]

Several East Los Angeles neighborhoods, made up mostly of Latino Americans, are located only a mile and downwind from the proposed hazardous-waste incinerator site. The Vernon incinerator was intended to be the "vanguard of the entire state program for the disposal of hazardous waste."[32] Residents of East Los Angeles questioned the selection of their community as host for the state's first hazardous-waste incinerator. Opponents of the incinerator saw the project as just another case of industry dumping on the Latino American community.

Mothers of East Los Angeles (MELA) led the opposition to the Vernon incinerator. MELA consisted of Latino American women who had originally organized against the state's plan to locate a prison in East Los Angeles.[33]

MELA targeted the South Coast Air Quality management District (AQMD), the California Department of Health Services (DHS), and the Environmental Protection Agency (EPA)—agencies responsible for awarding permits for the hazardous waste incinerator. MELA, like its South Central Los Angeles counterpart, was also able to garner allies to oppose the government-sanctioned hazardous-waste incinerator. MELA and its allies pressured CTTS through a lawsuit and the passage of a more stringent California state law requiring environmental impact reporting on hazardous-waste incinerators.

In 1988, as CTTS was about to start construction on the project, AQMD decided that the company should conduct environmental studies and redesign the original plans because of the new, more stringent state clean-air regulations. CTTS challenged the AQMD's decision up to the state supreme court and lost. In May, 1991, CTTS decided to withdraw because the lawsuits threatened to drive the costs beyond the $4 million the company had already spent on the project.[34] The incinerator was not built.

The other California community slated for a state-of-the-art hazardous-waste incinerator is Kettleman City, a rural farm-worker community of about twelve hundred residents. Because of their work, residents are exposed to dangerous pesticides. Moreover, the city is home to a Chemical Waste Management hazardous-waste landfill, California's largest hazardous-waste landfill.[35]

In 1991, the California Rural Legal Assistance Foundation, a public-interest law group, filed a class-action lawsuit, *El Pueblo Para el Aire y Agua Limpio (People for Clean Air and Water) v. County of Kings.* The lawsuit challenged the environmental-impact report in its use of English as the only language used to communicate risks to local residents when 40 percent of the residents speak only Spanish, and for its operating hazardous-waste incinerators in mostly minority communities. In 1992, a Superior Court judge overturned the Kings County board's approval of the incinerator, citing its impact on air quality and agriculture.[36]

THREAT TO NATIVE AMERICAN LANDS

As environmental regulations have become more stringent in recent years, Native American lands have become prime targets for garbage imperialism. Native American lands pose a special case for environmental protection.[37] Because of the special quasi-sovereign status of Indian nations, companies have attempted to skirt state regulations that are often tougher than the federal regulations.

The threat to Native American lands exists from Mohawk property in New York to Campos land in California.[38] More than three dozen reservations have been targets for landfills and incinerators. Nearly all these proposals have been defeated or are under review. In 1991, for example, the Choctaws in Philadelphia, Mississippi, defeated a plan to locate a 466-acre hazardous-waste landfill in their midst.[39] In the same year, a Connecticut-based company that had never operated a municipal landfill proposed to build a 6,000-acre municipal landfill on the Rosebud reservation in South Dakota. The project was later tagged "Dances with Garbage."[40] The Good Road Coalition, an alliance of grass-roots groups, led a successful campaign that derailed the proposal to build the giant municipal landfill on Sioux lands.

CONCLUSION

A new form of grass-roots environmental activism has emerged in the United States that emphasizes securing environmental justice for communities of color. Knowing that environmental racism is a major barrier to achieving environmental and economic justice for people of color, grass-roots activists have not limited their attacks to noxious facility siting and toxic contamination issues. Instead they have begun to seek change in destructive industrial production processes, wasteful consumptive behavior, urban land use and transportation, spatial housing patterns and residential segregation, redlining, and other environmental problems that threaten public safety.

People-of-color groups have begun to build a national movement for environmental justice. However, a national policy is needed to address environmental problems that disproportionately affect people-of-color, working class, and low-income communities. All communities deserve to be protected from the ravages of pollution. No one segment of society should have to bear a disparate burden of the rest of society's environmental problems.

Finally, pushing "risky" technologies and "dirty" industries off on people as a form of economic development is not a solution to the underdevelopment in impoverished Third World–like communities in this country and in similar communities around the world. Social-justice and equity goals must be incorporated into all levels of environmental decision making and policy formulation.

NOTES

1. See Robert D. Bullard, *Dumping in Dixie: Race, Class, and Environmental Quality* (Boulder, Colo.: Westview Press, 1990).
2. See J. A. Kushner, *Apartheid in America: An Historical and Legal Analysis of Contemporary Racial Segregation in the United States* (Frederick, Md.: Associated Faculty Press, 1980); Robert D. Bullard and Joe R. Feagin, "Racism and the City," in M. Gottiender and C. V. Pickvance, eds., *Urban Life in Transition* (Newbury Park, Calif.: Sage, 1991), 55–76.
3. Bunyan Bryant and Paul Mohai, eds., *Race and the Incidence of Environmental Hazards: A Time for Discourse* (Boulder, Colo.: Westview Press, 1993).
4. See Myrick A. Freeman, "The Distribution of Environmental Quality," in Allen V. Kneese and Blair T. Bower, eds., *Environmental Quality Analysis* (Baltimore: Johns Hopkins University Press, 1971); and Michel Gelobter, "The Distribution of Air Pollution by Income and Race" (paper presented at the Second Symposium on Social Science in Resource Management, Urbana, Ill., June 1988).
5. Patrick C. West, J. Mark Fly, and Robert Marans, "Minority Anglers and Toxic Fish Consumption: Evidence from a State-Wide Survey in Michigan," in Bryant and Mohai, *Race and the Incidence of Environmental Hazards.*
6. Robert D. Bullard, "Solid Waste Sites and the Black Houston Community," *Sociological Inquiry* 53 (Spring 1983): 273–288; and Robert D. Bullard, *Invisible Houston: The Black Experience in Boom and Bust* (College Station, Tex.: Texas A&M University Press, 1987).
7. United Church of Christ Commission for Racial Justice, *Toxic Wastes and Race in the United States: A National Study of the Racial and Socioeconomic Characteristics of Communities with Hazardous Waste Sites* (New York: United Church of Christ, 1987); Paul Mohai and Bunyan Bryant, "Environmental Racism: Reviewing the Evidence," in Bryant and Mohai, *Race and the Incidence of Environmental Hazards.*
8. Agency for Toxic Substances Disease Registry, *The Nature and Extent of Lead Poisoning in Children in the United States: A Report to Congress.* Atlanta: U.S. Department of Health and Human Resources, 1988, pp. 1-12.
9. Ibid.
10. Ibid.

11. Bullard, *Dumping in Dixie*, 7; Gerald Jaynes and Robin M. Williams, *A Common Destiny: Blacks and the American Society* (Washington, D.C.: National Academy Press, 1989), 144–145.
12. See Nancy Denton and Douglas Massey, "Residential Segregation of Blacks, Hispanics, and Asians by Socioeconomic Status and Generation," *Social Science Quarterly* 69 (1988): 797–817; and Robert D. Bullard, "Endangered Environs: The Price of Unplanned Growth in Boomtown Houston," *The California Sociologist* 7 (Summer 1984): 84–102; Bullard, *Dumping in Dixie.*
13. Denton and Massey, "Residential Segregation of Blacks," 814.
14. See Paul Ong and Evelyn Blumenberg, "Race and Environmentalism," Graduate School of Architecture and Urban Planning, UCLA (unpublished paper, March 14, 1990): 9; and Eric Mann, *L.A.'s Lethal Air: New Strategies for Policy, Organizing, and Action* (Los Angeles: Labor/Community Strategy Center, 1991), 31.
15. Joe T. Darden, "The Status of Urban Blacks: 25 Years after the Civil Rights Act of 1964," *Sociology and Social Research* 73 (1989): 160–73; Robert D. Bullard, "Solid Waste Sites and the Black Houston Community," *Sociological Inquiry* 53 (Spring 1983): 273–288; Joe R. Feagin, *Free Enterprise City: Houston in Political and Economic Perspective* (New Brunswick, N.J.: Rutgers University Press, 1987); and Robert D. Bullard, ed., *In Search of the New South: The Black Urban Experience in the 1970s and 1980s* (Tuscaloosa, Ala.: University of Alabama Press, 1989).
16. Commission for Racial Justice, *Toxic Wastes and Race*, pp. xiii–xiv.
17. Ibid., 18–19.
18. Pat Costner and Joe Thornton, *Playing with Fire* (Washington, D.C.: Greenpeace, 1990), 48–49.
19. Michael R. Greenberg and Richard F. Anderson, *Hazardous Waste Sites: The Credibility Gap* (New Brunswick, N.J.: Rutgers University Center for Urban Policy Research, 1984), 158–159; and Bullard, *Dumping in Dixie*, 4–5.
20. Bullard, "Solid Waste Sites" 273–288.
21. "Landfill Didn't Violate Equal Protection," *National Law Journal* (22 July, 1991): 28.
22. See Bullard, *Dumping in Dixie*, chapter 1.
23. U.S. Environmental Protection Agency, "Report of the Dallas Area Lead Assessment Study," (Dallas, Tex.: U.S. Environmental Protection Agency Region VI, 1983), 8.
24. Jonathan Lash, Katherine Gillman, and David Sheridan, *A Season of Spoils: The Reagan Administration's Attack on the Environment* (New York: Pantheon Books, 1984), 135–136.
25. Randy Lee Loftis, "Louisiana OKs Dumping of Tainted Soil," *Dallas Morning News*, February 12, 1992, A1 and A30.
26. U.S. General Accounting Office, *Siting of Hazardous Waste Landfills and Their Correlation with Racial and Economic Status of Surrounding Communities* (Washington, D.C.: U.S. Government Printing Office, 1983), 1.
27. Jane Kay, "Fighting Toxic Racism: L.A.'s Minority Neighborhood is the 'Dirtiest' in the State," *San Francisco Examiner*, April 7, 1991, A1.
28. For a detailed discussion of the incinerator controversy, see Louis Blumberg and Robert Gottlieb, *War on Waste: Can America Win Its Battle with Garbage?* (Washington, D.C.: Island Press, 1989), 155–188; and Dick Russell, "Environmentalism Racism," *The Amicus Journal* (Spring 1989): 22–32.

29. Jesús Sanchez, "The Environment: Whose Movement?" *California Tomorrow* 3 (Fall 1988): 11–17; Blumberg and Gottlieb, *War on Waste*, 168; and Russell, "Environmental Racism," 22–32.
30. Russell, "Environmental Racism," 28–29.
31. Russell, "Environmental Racism," 22–32.
32. Maura Dolan, "Toxic Waste Incinerator Bid Abandoned," *Los Angeles Times*, 24 May, 1991.
33. Mary Pardo, "Mexican American Women Grassroots Community Activists: Mothers of East Los Angeles," *Frontiers: A Journal of Women Studies* 1 (1990): 1–6.
34. Dolan, "Toxic Waste Incinerator Bid Abandoned."
35. For a detailed account of this conflict, see Julia Flynn Siler, "Environmental Racism? It Could Be a Messy Fight," *Business Week*, 20 May, 1991, 116; and Miles Corwin, "Unusual Allies Fight Waste Incinerator," *Los Angeles Times*, 24 February, 1991, A1 and A36.
36. "Judge Overturns Approval of Commercial Waste Incinerator," *Los Angeles Times*, 1 January, 1992.
37. Marjane Ambler, "The Lands the Feds Forgot," *Sierra* (May/June 1989): 44–48.
38. Ward Churchill and Winona LaDuke, "Native America: The Political Economy of Radioactive Colonialism," *Insurgent Sociologist* 13 (Spring, 1983): 51–68; and Conger Beasley, Jr., "Of Pollution and Poverty: Deadly Threat on Native Lands," *Buzzworm* 2 (September/October 1990): 39–45.
39. Adam Nossiter, "Proposed Toxic Waste Dump Divides Choctaws, Alarm Environmentalists," *Atlanta Journal-Constitution*, 5 February, 1991.
40. Thomas Daschle, "Dances with Garbage," *Christian Science Monitor*, 14 February, 1991.

3

Building a New Vision

Feminist, Green Socialism

Mary Mellor

MODERN INDUSTRIAL SOCIETIES offer two political visions: capitalism and socialism. The former claims that, with private ownership of natural resources and with human labor and consumption subject to individualized decision making, human society can maximize its wealth for all. The latter argues that maximum benefit is only possible by common ownership and collective decision making. The capitalist argument prevails as the basis of a "new world order," and the dreams of socialists have bitten the dust of Chernobyl and Tiananmen Square.

In place of the socialist challenge to capitalism, new social-justice movements have arisen: women's struggle against heterosexism and patriarchy, African American struggles against racism, anti-imperialist struggles in the South, and the green campaign for the planet. While all of these movements are vital, they do not replace the traditional struggle between the theory and practice of possessive individualism and the necessity of collective action for a just and equal society.[1] But one should not expect one last heave by revolutionary or reformist socialists who aim to capture the means of production or taxable benefits of the wealth-producing sector for the workers or the people. Women, racism, the planet or anticolonialism cannot be added on to the flawed base of industrial socialism. Industrial socialism has shared too many of the same characteristics as industrial capitalism, in particular the potential for unlimited material development and a male-dominated concept of the economic. We must break the boundaries of economistic thinking if we are to find a path to a sustainable, egalitarian alternative future.[2]

This chapter is based on my book *Breaking the Boundaries: Towards a Feminist, Green Socialism.* (London: Virago, 1992). I am grateful to Virago for permission to draw on the text here.

I argue for a feminist, green socialism, not because the feminist or green movement is more important than Black or anticolonial struggles or for that matter workers' struggles. Workers are increasingly female, Black, and from the South.[3] As transnational corporations scour the globe looking for new markets, tax havens, cheap labor and resources, and few pollution controls, one finds few distinctions among struggles concerning gender, racism, environment, production and local self-determination. Women and the natural world do, however, offer a fundamental challenge to both the material and ideological bases of capitalist production: Both lie outside the traditional conception of the economic. Until the advent of the green movement, the planet was treated as an externality,[4] while women, even in industrialized societies, live substantial parts of their lives outside the formal economy.[5]

SOCIALISM OR SURVIVALISM?

The case for retaining a commitment to socialism—although defined by feminism and ecology, in the first instance, rather than workers' struggles—is that the choice between individualism and collectivism is even more stark in the face of natural limits to human development. Even if only part of the ecological warnings of the greens is true, those who already suffer from discrimination, domination, exploitation, oppression, and exclusion from the wealth and resources of the planet will immeasurably worsen their burdens.

The benefits of the planet have been exploited by relatively few members of the human community, and there is no reason why such a deeply unequal world system would respond to an ecological crisis on the basis of social justice. Although in practice socialism has gone up many viciously corrupt blind alleys, the center of socialist philosophy contains a fundamental vision of a just and equal society, a vision that any progressive green politics must embrace.[6] If we are to build a sustainable egalitarian society in a world that has overreached its natural boundaries, greens should stand on the shoulders of socialists, not on their necks. Despite the United Nations' assertion of the close connection between economic and ecological inequality, the rich nations of the world have been unwilling to respond to these arguments, particularly at the hugely disappointing 1992 Earth Summit.[7] Without a clear commitment to social justice, only one alternative seems plausible: an uncaring, unjust, and unequal battle for survival.

The billion or so people who have benefited most fully from the plundering of the planet are predominantly white, male, and affluent. Ecological faultlines follow structures of economic power: from white to people of color, men to women, rich to poor, North to South. At the other end of those faultlines are the billion or so people who live on the boundaries of existence. The eco-economic chasm between rich and poor is created by a global market system that allows the overconsuming rich, white North, now joined by Japan, to scoop out the raw materials of the poor South, exploit its people as cheap labor, and use it as a dump for toxic waste. The South is locked into an economic system driven by the North without being able to find any mechanism with which to control and influence it. Even as the North extracts a surplus of fifty billion dollars from the South in debt repayments, it demands that the South slows its own development to provide the green "lungs" the planet needs to absorb the carbon emissions of the North. The North now has an ecological as well as an economic interest in maintaining the South's underdevelopment.[8]

Although the green movement places the global ecological crisis firmly on the global political agenda, the Earth Summit revealed the intractability of the current political situation. The rich nations are unwilling to threaten industrialism's culture of contentment, even at the cost of the future viability of human existence.[9] An ecological crisis leading to a limit on consumption in a world driven by the values of individual, commercial, and national economic self-interest can only bring one result: survivalism. Each person, group, organization, or nation will mercilessly struggle for its own continued existence. The small gains toward international cooperation, the elimination of absolute poverty, or a foundation for women's economic independence will be lost.

Struggles within the world market for raw materials suggest that capitalist economic growth and wealth do not necessarily relate to a country's riches in resources, thereby increasing the potential for international conflict. We have already seen eco-wars in the Middle East over the control of oil and water. Pressure on resources will also increase ecological and economic migration by those whose land and livelihood have been destroyed. This is already happening in both Europe and America, where migrant workers bear the brunt of a racist and nationalistic, if not neofascist, backlash. Deep ecologists fuel these situations by arguing against the misplaced humanism of those who oppose anti-immigration policies and call for an intensification of policies of population control.[10] The antihumanism of deep ecology has been widely criticized.[11] In a class-ridden, sexist, and racist culture, any discussion of "carrying capacity" or population must feed the prejudices of dominant groups.

If we are to sustain the future of the planet without risking a neo-Malthusian attack on already disadvantaged peoples or feeding the national self-interest of the richer nations, the green case for sustainability must be linked with campaigns for global social justice. The political collapse of industrial socialism places the future of the planet in the hands of a capitalist market economy united with other powerful forces—feudalism, patriarchy, colonialism, imperialism, militarism, and racism—to form a monstrous global structure of economic, cultural, and political power. Just abandoning industrialism, as some greens suggest, will not end the torment of the planet or its peoples. The necessary deconstruction of larger institutional systems will be long and hard.

An environmentally just system would not pile up privileges from the earth's exploitation for one part of humanity with the rest bearing the costs. Instead, both costs and benefits would be evenly shared within sustainable limits. The ecological crisis creates the conditions for a reassertion of the moral values of the common good against the self-interested divisions of the nation-state and the marketplace. Our choice is clear: a collective recognition of ourselves as a global human society with a commitment to ecological sustainability on the basis of social justice, or a globalized survivalism. The way forward lies with those who live outside of the economic system, women (particularly those in subsistence economies), and the planet itself.

THE END OF NATURE AND THE LIMITS OF CAPITALISM

Capitalists present the market economy as a game that anyone can play and one that, in the end, no one will lose, although some may win deservedly more than others. Only laziness and ignorance of those not willing to play the game, work hard, save, and take risks limit the system. Poverty is the result of a lack of development. This argument has paralyzed socialism, East and West. For Marxist socialists, poverty was an artifact of the capitalist system. Capitalism needed economic inequality and the dread of the poorhouse or unemployment to keep workers disciplined. After the revolution, the machinery of capitalism could be harnessed for the workers, "from each according to their ability, to each according to their need." Socialism would mean "leveling people up," not down. In making this claim, socialists have been locked into competing with capitalism in the promise of a better life, a race they were bound to lose. Capitalism could suck in resources from all over the world to feed its favored markets with no social or political obligation to those whose resources it had plundered or whose workers' health it had ruined.

When socialists challenged capitalism, pointing to its periodic economic crises and its deep inequalities, capitalism could ignore all criticisms with its eternal promise, growth. Those who owned and controlled economic resources justified their privilege by claiming that the exploited and impoverished would get their turn when development succeeded. The message from the green movement is that this will never happen. Instead, the global economy continues to sink.

Where socialism failed to puncture the capitalist illusion, the ecological crisis leads to questioning the mythic power of the market.[12] In a system with natural limits to production and consumption, profits cannot be conjured out of thin air. Nothing is without cost.[13] In a limited system capitalism can no longer conceal itself as a system for accumulating rather than distributing wealth. And without the promise of constant growth no moral basis exists for the private accumulation of wealth. The world capitalist economy has not proved that it will serve everyone adequately. Quite the opposite: Some people and countries are rich because other people and countries are poor.

For the South, global capitalism brings the mining of natural resources without industrial development, a growing circulation of capital with no capitalist transformation of the economy.[14] Even the World Bank, a major sponsor of export-led growth, admitted in 1990 that the market solution failed for poor countries and proposed reducing financial support to the commercial sector and redirecting it toward public-sector, labor-intensive employment and more finance for basic services such as health, education, and nutrition. In the North, economic growth has been largely an illusion where the quality of life deteriorates even as material standards increase.[15]

Traditionally, socialists have been wary of arguments about the limits to growth.[16] They argue that, far from encouraging the transfer of wealth from the rich to the poor, the rich will claim that everyone cannot be brought up to their level, so one should not even try. This confuses needs and wants. Certainly the whole world cannot rise to the standard of the average American whose lifestyle is already unsustainable. But no financial reasons explain why the peoples of the South should continue to live in poverty or ecological devastation.

The trillion dollars a year the world spends on arms would wipe out nearly the entire Third-World debt. It would take only $9 billion a year to secure the world's topsoil, $3 billion to restore the forests, $4 billion to halt desertification, $18 billion to provide readily available contraception worldwide, and $30 billion for clean water.[17] Commitment of resources to military expenditure has meant that nuclear missiles could travel from Europe to Moscow in minutes, while a woman in Africa had to walk for several hours a day to fetch water.[18] These priorities are not the neutral decisions of a market, they are the priorities of powerful men in powerful nations, men whose gender, race, and class interests drive the capitalist system and its worldwide system of accumulation and deprivation.[19] The challenge to the global market economy comes not only from the ecological crisis but from women who live much of their lives outside its framework.

WOMEN BEYOND THE MARKET

It is not only the poor who have been excluded from the market bandwagon. Women, even in prosperous countries, live largely outside of the formal economy. The definition of valued work in the market has been defined not just by capital but by men.[20] A major weakness of industrial socialism is its attachment to the productive worker, a concept that locks it within the boundaries of an economic system defined by men and capital. Consequently, women who only do unpaid domestic work in market economies have no independent right to resources other than through a male wage or welfare system. Even with a social wage, this system depends on the wealth created by the male, capitalist economy. Yet the large number of hours women spend in domestic and subsistence labor outside of the marketplace is a vast resource to societies all over the world.[21] Women are also responsible for securing the emotional needs of those around them and for maintaining the functioning of the local community by minimizing conflict in human relationships.[22]

Saying this does not imply that women are innately superior to men, that they are more caring, loving, nurturing, or self-sacrificing. Quite the opposite is true: Male domination forces women into economically unrewarded invisibility.[23] Across the globe women engage in a variety of tasks, from agriculture to industry, from water carrying to teaching, for little reward. Women are "managers of human welfare," meeting survival and subsistence needs.[24] If all else fails, they sell their own bodies to feed their families or agree to be sterilized for money or food.[25] Most of women's work is unpaid because it is never bought or sold as a commodity on the market. Crops grown, meals prepared, or clothes made are used directly, so that women do not produce a tangible product other than a healthy child, a well-cooked meal, or a few gallons of water fetched from a distant well. On this basis most of women's work counts, quite literally, for nothing.[26]

The invisibility of women and women's work affects world decision-making systems so that aid programs often ignore women's needs for access to clean water or sanitation.[27] The United Nations System of National Accounts (UNSNA), for example, explicitly excludes women's subsistence work from its definition of production, arguing that such work has no value "since primary production and the consumption of their own produce by nonprimary producers is of little or no importance."[28] Not taking account of women's subsistence work has profound consequences for countries with large subsistence economies, since their ability to benefit from aid or loans is calculated on cash incomes rather than the real productive capacity of the country.

The invisibility of women is not confined to subsistence economies. Marilyn Waring, as a member of the New Zealand government, found it very difficult to prove that child-care facilities were needed by women defined as "inactive" and "unoccupied."[29] This does not mean that the solution to the environment or women's work is necessarily to bring them within market values. If resources are priced, they will go to the highest bidder; by the year 2000, multinational companies will control more than 50 percent of the exploitable resources of the world.[30] If women's work is given a wage, it may be very low, reflecting the low esteem in which a sexist culture holds her work. For Vandana Shiva, women are a living indictment of the market system: "Indian women...have challenged the western concept of economics as production of profits and capital accumulation with their own concept of economics as production of sustenance and needs satisfaction."[31]

NATURAL WORLD AS COMMOMWEALTH

From an ecological perspective, any division of the global ecosystem is unnatural. The revolutionary potential of the ecological crisis lies in its challenge to the boundaries and divisions within and between societies. Private ownership and profit-oriented economies within a system of nation-states are not conducive to seeing the natural heritage of the planet as a common resource for all humanity. The resources of the earth are not experienced as a "common," in the traditional sense of land and resources belonging to a community, as a source of wealth for all and the responsibility of all. Instead, society values them only as money, what Marx called the "fetishism of commodities." Marx claimed that seeing the products of human labor only as commodities ignores the real relationship of exploitation between human and human. The same is true of the relationship between humanity and nature. The costs of production become an externality, a cost borne by the planet and by those people who suffer the environmental consequences. This situation is made worse by consuming products away from the site of both economic and ecological exploitation.

The commodified market economy quickly turns natural resources into cash. The seas are stripped of their fish, the hillsides of their trees, the rivers of their fresh water, and the land of its fertility. Under the logic of self-interest and survivalism, nothing will stop the fisher fishing the last fish, the logging company taking the last tree, or the dam withholding the last drop of water.

The classic analysis of "fishing the last fish" or, rather, grazing the last blade of grass, is the crux of Garrett Hardin's "tragedy of the commons."[32] Hardin based his influential article on a parable set forth in a neo-Malthusian pamphlet issued in 1833. The parable hypothesizes a shared common grazing land with a finite carrying capacity used by local herders to graze their cattle. On the assumption that the herders follow their own individual self-interest, each attempts to maximize self-interest by trying to graze as many cattle as possible, thus destroying the resource for everyone. According to the parable, each herder is aware of the damage overgrazing will do but still persists because it is in each person's self-interest. The net economic benefit to themselves of grazing one more head of cattle is less than the cost each bears for the decline of the common, which is shared with all the other herders. Hardin's solution is to privatize the commons or impose some kind of authoritarian control.

The fallacy in the "tragedy of the commons" parable lies in the concept. Common land assumes a community, something held in common. However, Hardin's parable assumes that the herders behave according to the principles of the individualized marketplace, on the basis of economic self-interest and possessive individualism. He does not consider the possibility that the common could be administered collectively, fixing a maximum number of cattle per herder. Certainly, people pushed by need into marginal land all over the world are destroying their habitat, but the parable describes greed, not need. If the motivation is personal greed, and the grazing land is treated as an ownerless resource, then nothing will stop a herder grazing the last blade of grass. Hardin's logic holds only if the grazing land is *not* being treated as a common. The concept of a common is social by definition; it is understood as a boundaried, limited resource.

The tragedy of the commons occurs not just because each individual tries to secure their self-interest but because they know or fear that everyone else is doing the same. It is not the presence of individual self-interest but the absence of a sense of commonality that is most crucial. If we are to prevent someone fishing the last fish, they must be assured that no one else will do so. We must secure the commons, that is, those resources essential for our mutual sustainability. The only way that the commons can be secured is to persuade individuals that their needs will be met, and they need not grab and hoard for themselves. Security of resources cannot be separated from a commitment to equity in securing basic needs.

Undoubtedly, we face increasing scarcity of clean air, water, raw materials, energy, food, and space. We can either meet this future as individuals driven by need or greed or as a political community based upon sharing and mutual support. The only political philosophy that can effectively secure the commons by recognizing the principles of equality and mutuality in securing basic needs is a revitalized socialism.

MAKING IT HAPPEN

Progressive political and social change requires two things: a movement that has captured the public imagination and a faltering structure of domination. A movement that draws on the insights of green, feminist, and socialist ideas can provide that vision. Despite its strutting power, the new world order is faltering. It cannot solve the problems of either global poverty or a global ecological crisis. The first may not matter to the rich, but the second does. The crisis of the environment is global, and the rich cannot hide for long from toxins, ozone depletion or global warming. The campaigns for public health in Britain in the nineteenth century are a good analogy. The rich were convinced not by the health needs of the poor but by their inability to insulate themselves from the germs that assailed the poor.

A feminist, green, socialist vision would embrace the needs of the planet and all its species, including humanity. It would start from the perspective of those who are already suffering through racism, sexism, poverty, and exploitation, with a minimal utopia of a basic standard of living for all humanity. This would be green, because a commitment to basic needs represents a move to a sustainable society; it would be socialist as the first step towards an egalitarian society; it would be feminist, because, particularly for poor women, "the fulfillment of basic needs is the priority issue."[33]

The challenge of feminist, green socialism to the new world order is that politics must replace economics as its main determinant. The minimum standard of a dignified human life for all can no longer be the by-product of an economic system; it must be the central pivot of political life. Basic needs are not just physical; they are also emotional and intellectual. Manfred Max-Neef has identified nine basic needs: subsistence, protection, affection, understanding, participation, leisure, creation, identity or meaning, and freedom.[34] A society in which people are mutually responsible for providing all of these basic needs for each other would necessarily be feminist as it must transcend the present sexual division of labor. Emotional support, for example, would no longer be the major responsibility of women, and opportunity for leisure and creativity would be welcomed by those who are now working the double shift of paid and unpaid work. It would also satisfy green demands for a displacement of material values by nonmaterial values.

In material terms, the basis for ecological sustainability and the reestablishment of a basic-needs economy is very much greater in the South than in the North. In the South many people, particularly women, are still connected with the earth and retain their old skills and knowledge. The most important problems are the incursion of the capitalist agribusinesses with a consequent exclusion of poor women and men from land ownership and the loss of plant and animal diversity through monoculture.

For the majority of the world's peoples, particularly women, land rights are vital to environmental justice. The enormity of this struggle cannot be overestimated, and the choice between individual and collective principles of ownership is crucial. Women's access to land is generally based on their common rights as members of the community. If individual ownership is established, this usually rests with the male head of the household, and women's independent right to land is lost. Individual ownership also carries the risk of indebtedness and land grabbing by wealthier landowners. Collective models of land ownership are essential if womens' rights are to be retained and expropriation by indigenous or external landlords prevented.

In the North, knowledge, land, and biodiversity have already gone. While economic benefits are much greater than in the South, real wealth is virtually nil, at least for the average person. Most people have skills that can only be sold as labor. Consumption, from prescription medicine and prepacked food to entertainment, is almost totally commodified. Expectations and diets are geared to economic colonialism. To escape such a massive level of commodification will mean reclaiming and relearning lost skills, particularly in health, horticulture, and appropriate technology.

In the commodified economies of the North, it may seem a Herculean task to build an alternative to the present economic system, but it has already been done. In the early nineteenth century British workers, themselves only recently displaced from the land, were faced with poor food and exorbitant prices. In protest they formed a cooperative movement, buying, producing, and distributing their own food and other goods. The cooperative movement grew into a massive organization covering not only retailing but production, housing, banking, and insurance. By the 1950s, twelve million Britons were members of cooperative societies, a major employer and channel of mobility for working class people with its own education system. Cooperative societies formed, and still form, a democratically controlled, noncapitalist organization linked to a worldwide cooperative movement.

Although the cooperative movement has not been without its problems—like many other organizations, it is male-dominated—it seems to be holding its own. In Britain, despite the recession, consumer cooperatives gained a market share in 1991 and 1992. It is also one of the biggest landowners in Britain. Women are also playing an increasingly central role in recent cooperative movements, most notably in the Japanese Seikatsu movement, which has grown from a milk-buying cooperative of two hundred women in Tokyo in 1965 to a 170,000-strong movement providing organic foods and social welfare services. It lobbies on environmental and health matters, and many of its members have been elected to their local councils. The overall aim of the Seikatsu movement is to create an alternative way of life based on the principles of self-government and ecological sustainability.

Worldwide the cooperative movement provides a massive network that could directly connect the commodified consumers of the North with the subsistence farmers of the South, bypassing the capitalist market system. Worker cooperatives could also form the basis of local employment and trading schemes, provided they were part of a wider movement for social change and not isolated as economic units.[35]

Creating a vision is about breaking the boundaries that constrain us. Structures of domination exist not only in reality but in our minds. Militarized nation-states and multinational conglomerates cannot be thought away, but we can liberate our thinking from their forms of reasoning. Breaking through the boundaries they have created in our minds will empower us to begin the struggle to confront them. That struggle must not be fractured by the unnecessary divisions of reform vs. revolution, personal change vs. political struggle, ideological purity vs. well-meaning confusion, or the sustainability of the planet vs. the livelihood of workers. What matters is that we build "bridges of power" reaching out to each other and moving toward a common dream.[36]

NOTES

1. The classic critique of possessive individualism is C.B. Macpherson, *The Political Theory of Possessive Individualism* (Oxford, England: Oxford Univ. Press, 1962).
2. In this I am following the pioneering work of Hazel Henderson. For a fuller development of her arguments, see *Creating Alternative Futures* (New York: Perigee Books, 1980), and *The Politics of the Solar Age* (New York: Knowledge Systems, Inc., 1988).
3. See Maria Mies, *Patriarchy and Accumulation on a World Scale* (London: Zed Press, 1986), and Swasti Mitter, *Common Fate, Common Bond* (London: Pluto, 1986).
4. See K. W. Kapp, *The Social Cost of Business Enterprise* (Nottingham, England: Spokesman, 1978).
5. See Heather Jon Maroney and Meg Luxton, eds., *Feminism and Political Economy: Women's Work and Women's Struggles* (London: Methuen, 1987).
6. This point has been made repeatedly by eco-socialists and eco-anarchists such as Andre Gorz, *Ecology as Politics* (Boston: Beacon Press, 1980); Martin Ryle, *Ecology and Socialism* (London: Radius, 1988); and Murray Bookchin, *Remaking Society* (Montreal: Black Rose Books, 1989).
7. See World Commission on Environment and Development, *Our Common Future* (Oxford, England: Oxford Univ. Press, 1987).
8. See James O'Connor, "Uneven and Combined Development and Ecological Crisis: A Theoretical Introduction," *Race and Class* 30 (1989): 1–11.
9. See J. K. Galbraith, *The Culture of Contentment* (London: Sinclair-Stevenson, 1992).
10. Bill Devall, *Simple in Means, Rich in Ends* (London: Green Print, 1990), 189.
11. One of the major ciritics of this tendency in deep ecology is Murray Bookchin. For a broader treatment of his argument, see *Green Perspectives* (Burlington: Vt.: Green Programme Project, 1988).
12. See Jeremy Seabrook, *The Myth of the Market* (Bideford, England: Green Books, 1990).
13. See Barry Commoner, *The Closing Circle* (London: Jonathan Cape, 1972).

14. See N. Shamugaratnam, "Development and Environment: A View from the South," *Race and Class* 30 (1989): 13–30.
15. See Richard Douthwaite, *The Growth Illusion* (Bideford, England: Green Books, 1992).
16. See Joe Weston, ed., *Red and Green* (London: Pluto, 1986).
17. See Norman Myers, *The Gaia Atlas of Future Worlds* (London: Robertson McCarta, 1990).
18. See Gita Sen and Caren Grown, *Development Crises and Alternative Visions* (New York: Monthly Review Press, 1987).
19. See Immanual Wallerstein, *Historical Capitalism* (London: Verso, 1983).
20. See Pat Armstrong and Hugh Armstrong, "Taking Women into Account," in *Feminization of the Labour Force,* edited by Jane Jenson, et al. (Cambridge, England: Polity, 1988).
21. See *The New Internationalist* 181 (March 1988).
22. See Nicky James, "Emotional Labour: Skill and Work in the Social Regulation of Feelings," *Sociological Review* 37 (1989): 15–42.
23. Mellor, *Breaking the Boundaries,* 252–255.
24. Sen and Grown, *Development Crises,* 8.
25. Asian and Pacific Women's Resource Collection Network, *Asian and Pacific Women's Resource and Action Series: Health* (Kuala Lumpur, Malaysia: Asian and Pacific Development Centre, 1989).
26. Marilyn Waring, *If Women Counted* (London: Macmillan, 1989).
27. See Irene Dankelman and Joan Davidson, *Women and Environment in the Third World* (London: Earthscan, 1988), and Brinda Rao, "Struggling for Production Conditions and Producing Conditions of Emancipation: Women and Water in Rural Maharashtra," *Capitalism Nature Socialism* 2 (Summer 1989).
28. Waring, *If Women Counted.*
29. Ibid., 2.
30. Jeremy Rifkin, *Time Wars* (New York: Hold and Co., 1987).
31. Vandana Shiva, *Staying Alive* (London: Zed Press, 1989), xvii.
32. Garrett Hardin, "The Tragedy of the Commons," *Science* 162, 12438.
33. Sen and Grown, *Development Crisis.*
34. Manfred Max-Neef, "Human Scale Economics, the Challenges Ahead," in *The Living Economy,* ed. Paul Ekins (London: Routledge & Kegan Paul, 1986).
35. Mary Mellor, Janet Hannah, and John Stirling, *Worker Cooperatives in Theory and Practice* (Milton Keynes, England: Open University Press, 1988).
36. Lisa Albrecht and Rose M. Brewer, *Bridges of Power: Women's Multicultural Alliances* (Philadelphia: New Society, 1990).

4

The Promise of Environmental Democracy

John O'Connor

U.S. ENVIRONMENTAL POLICIES are not working. Global warming, ozone depletion, species extinction, rain forest destruction, desertification, and the contamination of our land, water, and air by toxic chemicals, oil spills, radiation leaks, and pesticide use continue largely unabated. Despite the volumes of federal regulations and laws created over the past twenty-five years to protect the environment and public health, conditions have not improved. What makes our policies and regulations fundamentally ineffective? What ailment afflicts American democracy, rendering a concerned populace powerless against the environmental threat? Is there a way to revitalize our society and restore our environment?

I argue that democratic ideas that were part of the founding of the United States are critical to modern environmentalism; the promise of American democracy's rule of the people has been subverted by the rule of the few; environmental laws have been designed to fail in order to protect the financial interests of the world's largest polluting industries; and the best hope for preventing the destruction of the environment and public health is through the practice of environmental democracy, which opposes the power of environmentally-threatening corporations by reestablishing direct democracy and by reacquainting U.S. citizens with rights that are as old as the country itself.

THE DREAM OF THE FOUNDERS

Debates continue among scholars about the exact intent of the nation's founders. Did the Declaration of Independence and the ensuing American Revolution represent the progressive politics of the era with its proclamation of equal rights to all? Did the

U.S. Constitution represent a step backwards, putting the interests of large property holders ahead of the political rights of individuals? Was the propensity for unlimited growth and the domination of nature—ideas that run counter to ecological ideals—part of the country's foundation? Clearly slavery, the destruction of Native American nations, and the long-suppressed rights of women suggest caution in sorting out what we can draw on from the founders.

This article explores those elements of our early ideals of democracy that are most helpful in revitalizing both democracy and the environment. I argue that environmentalism is an extension of democratic rights to both nature and people. The history of U.S. democracy is a history of expanding rights, first to white men, then to African Americans, then to women. Recent environmental legislation (e.g.,The Endangered Species Act, Clean Air and Clean Water Acts) implicitly recognizes that a human being's rights to "life, liberty, and the pursuit of happiness" are impossible to exercise if society fails to preserve the integrity of the natural world. Environmentalism is deeply rooted in America's past.[1]

Where did these values come from? The founders drew from both European theory and from the ideas and practices of Native Americans. From this unique amalgam emerged three themes particularly relevant to the modern environmental debate but too often neglected: democracy, equality, and sustainability. Many historians now acknowledge the Native American roots of American democracy, even in the work of philosopher John Locke. The early colonists fashioned a political system partly borrowed from the Native American experience.[2]

Benjamin Franklin and Thomas Jefferson borrowed from Indian social organizations when seeking inspiration for a new political system. The Iroquois League (or Six-Nation Confederacy)—a near neighbor of the thirteen colonies—had almost two centuries of political unity prior to the American Revolution. Their constitution long predated the American version. Franklin attended negotiations between colonists and Iroquois leaders. He observed the Confederation's political practices, noting its provisions for impeachment, initiative, referendum, recall, personal liberties, rights of women, freedoms of speech and religion, principle of equality, division of power, and popular rule.[3] As early as 1754, Franklin had a plan for uniting the thirteen colonies, modeled upon Iroquois society, in which sovereignty would be shared between the individual states and a federal government. Jefferson, impressed with Native American political concepts, admired their decentralized political organization. There were no oppressive laws, yet daily life was lawful. Power was held by those whose natural authority was recognized by their fellows. Such people, gladly participating in the decisions that shaped their lives, were central to Jefferson's vision of a flourishing democracy. While a representative system of government was seen as a necessary expedient to extend the democratic ideal across a large territory, philosopher Hannah Arendt argues that Jefferson had "a foreboding of how dangerous it might be to allow a people a share in public power without providing them...with more public space than the ballot box."[4]

Jefferson believed that local and state institutions were basic elements for a successful national representative democracy. He proposed a political unit of direct democracy called the ward republic. This local institution never fully developed but finds its closest relative in the New England town meeting—perhaps the last vestige of direct democracy in the United States. Modeled after both the town meeting and the Native American tribal council,[5] Jefferson's ward republics were not intended to supplant federal, state, or county governments. Instead they brought politics down to a human scale—individual wards of several hundred people that every local citizen could attend. As active members of their ward, citizens could initiate proposals and perhaps watch them develop to the federal level, thus making every citizen a legislator. The federal government could not enact a major law until it brought the proposals for debate and decision to the ward level.

The nation's founders, for the most part, did not call for leveling, or a strict equal distribution, of wealth, but they took steps to ensure that the democratically elected government extended the nation's potential wealth to all. Popular will, not wealth, would shape society. Democracy was to be the great insurance policy covering the market place, insisting that it served the interests of the entire American community. Equality required all private wealth and power to be subject to the needs and welfare of the larger American community. When government failed in this mission, it was the right and duty of the citizenry to alter or abolish" that form of government and replace it with a more workable model.

Realizing that they could not foresee all future developments, America's founders created the Ninth Amendment, providing rights to disenfranchised individuals or groups as the American experiment in democracy unfolded. This long-established acceptance of new rights emerging as society develops is basic to empowering society's reversal of the environmental crisis.

American democracy was intended to survive. Both Native American and European cultures held that it was the responsibility of each generation to ensure its continued existence. Long before sustainability was a fashionable subject for debate, it was a natural political assumption to America's elder statesmen. The six Iroquois nations insisted that their decisions take into consideration the needs of their descendants seven generations into the future. But the early promise of American democracy is a fading dream.

THE DECLINE OF DEMOCRACY

As Americans entered the 1990s, the promise of democracy where the citizenry—either directly or through their representatives—controlled their own collective fate was becoming a distant memory. How had democracy departed so far from the vision of the founders that the needs of the people were neglected and their voice unheard? Alexis de Tocqueville, traveling through the United States in the 1830s, believed that the young country had found a political solution that ensured equality, as well as direct representative democracy, from the continuation of assembly-like town meetings and the representative system of state and federal government. He did express reservations, however, about the role industry might play in America's future. The growing separation between master and workman was an unhealthy trend. He had already witnessed the social and natural devastation that New England's new class of manufacturing barons had created, and he worried about how this decision would affect democratic practice.

De Tocqueville's foreboding of an industrial aristocracy has materialized on a grand scale. The vast difference between the wealth of the people who own and control the multinational corporations and the great majority of Americans makes a mockery of the principle of equality, a necessity to any democratic system. Democracy is undermined as its constitutionally mandated institutions have been captured by those whose sole motive is financial gain to the exclusion of environmental protection as well as the economic welfare of most Americans.

Today, multinational corporations undermine democracy through both their power to control information and their lavish spending on elections. General Electric, one of the nation's largest chemical producers and generators of hazardous waste, owns the National Broadcasting Company (NBC). In 1990 the largest oil and chemical firms spent $16 million defeating California's "Big Green" initiative, outspending environmentalists five to one.[6] Producing the right information in order to affect the outcome of legislation or regulatory proceedings may be just as important as giving money to candidates. Both require resources that preclude the involvement of average citizens.[7]

International corporations contribute millions of dollars to conservative think tanks such as the Heritage Foundation and to trade associations. Few citizens have the time or resources to track legislation, draft comments, and attend hearings. They certainly cannot afford to pay for the production of information and to hire lobbyists.[8]

ENVIRONMENTAL REGULATION: DESIGNED TO INCREASE THE POISONING

Despite nearly a quarter century of apparently vigorous government action and the creation of weighty new laws and agencies, the environment continues to deteriorate. Neither the weight of the regulatory programs, the post-1970 laws, or the common law guarantees the protection of the environment or of the public's health and safety. With a few notable exceptions, all acts of regulation accept an ever-increasing load of pollution into the environment, simply shifting the toxics or carbon dioxide from one route of exposure to the next. Even the best federal laws with broad international agreement, like an early phase-out of CFCs, has been unsuccessful in stopping the hole in the ozone layer from widening.

Our environmental legal system does not outlaw those forms of production that undermine the sustainability and life-giving forces of the physical environment. Pollution regulation accepts that our system of production creates pollution, and that our cars, factories, and farms use toxins and nonrenewable fuels.

For example, regulation does not challenge the production of synthetic pesticides. Instead, it accepts the widespread use of pesticides and subsequent environmental destruction and tries ineffectively to "control" their use. In the manufacturing of automobiles, regulation fails by not governing the type of engine and accepting the private choice of General Motors and Ford Motor Company to produce high-compression engines that are both fuel inefficient and environmentally destructive. Current law requires adding a device to the "end of the pipe"—the catalytic converter— to remove some but not all of the pollution.

In industrial manufacturing, the product and process is entirely the corporation's decision. Governments only influence what is done with the by-products (waste) of production. A system of environmental democracy can respect private ownership of production while requiring that manufacturers who threaten the environment and public health subject their decisions on what is produced, and how, to democratic review. Further, decision-making structures thoroughly grounded in a democratic process might move people to enact measures that really protect the environment. For example, a combination of electric engines, corn-fueled engines, and light rail might replace high-compression gasoline engines, if such decisions were subjected to democratic review.

Because the government does not normally challenge what is produced and how, it tends to regulate after the fact—a product or process is generally not subject to review until or unless large numbers of people are hurt, sickened, or killed. Then the government investigates and takes only protective measures. The histories of how the federal government handled asbestos, coal dust, lead, vinyl chloride, PCBs, DBCP, DDT, and many other significant public-health or environmental threats are good examples.

Dr. Barry Commoner shows that pollutants can be stopped from entering the environment by either pollution control or by eliminating the activity that generates pollutants (pollution prevention).[9] While the former typically results in waste shifting, and hence continued increases in amounts, only the latter promises to eliminate the source of pollution. Unfortunately, almost all of the government's resources are tied up in pollution control efforts that simply shift waste from one medium to another. The incineration of hazardous waste, for instance, transfers pollution via gases and toxic ash to other mediums—the air and back to the land and water—in a way that does not address the source of the problem: the continuing creation of toxic waste and dangerous emissions.

Finally, and perhaps most important, the regulatory system is fundamentally undemocratic. The decision makers are not the people, or even the people's representatives, but rather administrative law judges, engineers, technical experts, and private industry-bound bureaucrats who produce regulations inconsistent with the will of the people, and sometimes at odds with the letter of the law. In general, regulators are also not subject to popular forms of democratic review. The limited democratic process through Congressional representation that might have financial oversight of the Environmental Protection Agency cannot even ensure that executive-branch regulation is consistent with the law enacted by Congress. The Clean Air Act of 1990 and subsequent regulations offers dozens of examples of how regulations issued by the executive branch of the government thwart the will of the people. How can the system be made more democratic?

ENVIRONMENTAL DEMOCRACY: RESTORATION OF THE DREAM

Effective democracy requires changing the power relations between the citizenry and corporations and between the citizenry and the government. The centralization of political and economic power, as well as its increasing concentration, is at the heart of the ailment afflicting U.S. democracy. For the people to regain their sovereignty and restore their environment, fundamental changes in the structure of government are required that permit effective discussion, debate, and decision making. Every citizen, regardless of wealth, should be given an equal opportunity to bring ideas to stop environmental decline directly to the American people. In addition, each citizen will need new rights, in relation to polluting or poisonous industries, in order to allow citizens an opportunity to protect themselves and their communities from the everyday abuses of economic power, against which no level of regulation or conventional governance can fully protect them. The practice of new rights can provide a greater level of public health, safety, and environmental protection than existing law.

I begin with an examination of a proposal for a form of direct democracy and then discuss the rights individuals need in relation to corporations. These new rights, coupled with a new level of direct democracy, are fundamental to sustaining the environment and revitalizing the democratic character and spirit missing in so many Americans. The proposed model represents one piece of a larger strategy for social change to bring about environmental justice.

A NATION OF TOWN MEETINGS

In proposing a new design for democracy, representative institutions would be supplemented by new autonomous units. The aim is to revitalize the representative system through local democratic processes, allowing continual engagement against both political inequality and environmental deterioration.

Today's democratic system lacks the vital links between citizen and city hall, citizen and state house, citizen and federal government. The proposal is to add a political unit Jefferson called the "ward republic," similar to a New England town meeting, with perhaps two thousand adult members. Meetings held quarterly, or more often if one hundred members petitioned, would assemble to tackle the great issues of the day. This new level of direct democracy would be the ultimate people's check against the hitherto inaccessible forces of government. The ward assembly would have the power to propose or veto legislation not just locally but on a state and federal level when a majority of other wards debate and decide in favor of a veto or proposal. It could also establish agreements and contracts with corporations that were at least as effective as state or federal law but that would provide additional levels of protection for health, safety, and the environment.

How would it work? How could every citizen become, in essence, a legislator? Suppose a citizen had a great idea on how to protect the environment while at the same time improving the economy. How could a system of assemblies within wards enact this idea into law? First, the proposal would have to pass by a simple majority at the local ward level. Next, it would be debated and approved in at least in a majority of wards at the town, city, or county level. Then, a majority of the wards in a state would debate and decide the proposal. If a simple majority of wards in a state approved, the proposal would be voted on nationwide. A majority of states voting in favor, using the same ward process, would transform this once local initiative into federal law. The local ward republic would also have the right to insist that an industry in a neighboring community make changes necessary to improve conditions of health, safety, or the environment. The ward would have the power to require a collective bargaining agreement between itself and the industry in question. The ward concept is one option, and details would have to be worked out along the way since such a proposal would evolve considerably over time.

Robert A. Dahl, a preeminent scholar of democratic theory, adds another dimension to creating a stronger democracy: democratizing capitalism. In his two most recent works, *A Preface to Economic Democracy* and *Democracy and its Critics*, he argues that there is no unlimited "right of property," and even if there were such a right, it is secondary to the right of self-government.[10]

Moreover, because the private ownership of corporations impedes the democratic process by stifling the individual's voice and creating a lopsided balance of power, a state would be justified in taking what steps seemed necessary to democratize the economic arena, particularly the workplace. One necessary step towards environmentally sensitive democracy, beyond our proposal of ward assemblies, is that the federal government—particularly the Congress—must play an increasing role in deciding what is produced and how, if the products or processes in question threaten public health or the environment.

I agree with Dahl on the need to more fully democratize capitalism. I disagree that giving workers democratic participation in corporate policies will sufficiently address political inequality or the environmental crisis. Certainly workplace democracy would promote better trained and skilled citizens. But by narrowing our democratic voice to include only the firm in which one works, we are in danger of creating the nation of "factions," against which James Madison warned us in the *Federalist Papers*, where self-interest becomes selfish interests. A system of empowered assemblies within each political ward, in addition to the existing representative structure, offers each citizen an equal opportunity to exercise power all the way to the federal level. The voice that best captures the common good is heard only if our system of democracy allows all voices equal consideration, starting at the most local level.

I realize that these ideas raise as many questions as they answer. Will a new layer of direct democracy, even when coupled with the new rights discussed below, adequately restrain polluting corporations that operate across borders? Aren't these multinational corporations becoming independent of nation-states? Won't stronger antipollution provisions coming out of ward assemblies encourage the corporations to move elsewhere in the global economy? Can even the most outlaw enterprise be restrained by any government authority in the current climate of continuing centralization and concentration?

To these questions I have no easy answer other than that my proposals are part of a larger array of options that people must experiment with as we move towards solutions. I do not believe that anyone will find an absolute solution. Instead, my modest proposals speak to concrete mechanisms that create the public space to draw on the ideas of many more people. The proposals are capable of revitalizing an inactive citizenry. Again, many more minds is the point. I have democratic faith that an energized and engaged populace will eventually answer these difficult questions.

NEW RIGHTS

In Western civilization the concept of rights is basic to improving the workings of democracy. The right of life [11] is a precondition for all other rights, a sort of moral safety net that every government claiming to be just aspires to provide. But what does the right of life entail? Is it survival, or does one have a right to a minimum standard of living? Each successive generation must answer these questions. The implication is, however, that the right of life includes security from harm.

In coupling the right of life with liberty and the pursuit of happiness, America's founders clearly demonstrated their belief that there was more to living than mere subsistence. They also provided an effective means of combating the social and environmental devastation wrought by polluting industries. The right of life cannot be guaranteed if corporations can poison the soil, air, and water with cancer-causing chemicals. The pursuit of happiness is an empty promise for the six million children, predominantly poor, who may suffer learning disabilities because of permitted environmental lead levels. If life is a right, then we must protect those elements that sustain it. Since drinking and breathing are essential to life, then clean water and air are rights as well. What good are any rights—free speech, religion, or assembly—if we are not healthy enough or alive to enjoy them?

Those who pollute protest that they too have rights. Property is a right, they claim, and their use of it constitutes their own means of pursuing happiness. For generations this argument has stifled dissent, but the situation has become so urgent that environmentalists and community organizations are at last willing to disrupt this false notion by asserting that their health and that of their children—in other words, their right of life—has priority. Proprietors of poisonous capital do not have the sole right to decide how their property is enjoyed once their actions endanger the public.

Publicly invoking our most basic rights often disrupts the peace. But the effort is essential, and the results can be significant. Collective action by well-organized communities can not only force changes in the behavior of antisocial industries but can also result in the acceptance of new rights—at first in practice, later in law. This is compatible with the whole nature of the United State's evolving democracy and in harmony with the Ninth Amendment and with Jefferson's vision of rewriting laws to suit the needs of successive generations. It is one of the ways American democracy was designed to work. Rights are generally won, not granted. Neither the environmental movement today, the campaign for women's suffrage, or civil rights before that could wait for the codification of democratic rights. Now, like then, we must assert these rights until their moral weight becomes too compelling for lawmakers to resist.

Evidence suggests that the campaign to establish new environmental rights is succeeding. Since the late 1970s, thousands of communities, neighborhood groups, and workplace organizations have insisted that the proprietors of local plants provide information to workers and the public about the types, quantities, and health effects of toxic materials used or disposed of by the enterprise under their authority. The federal government accepted the principle of this claim in the federal Right to Know Law of 1986, which gives the public access to information about toxic substances that could affect their right to a healthy life. While imperfect, the law indicates that American democracy can still function if pressed by an organized citizenry, challenged though it is by an unjust concentration of economic and political power. When people organize efficiently to assert their rights, the law of the land will change. Right to know became state and federal law only after thousands of local community organizations and unions first asserted and won their rights locally.

As a result of many new grass-roots campaigns organized around toxics and public-health issues, in essence based on the right of life, three new rights are emerging that alter the relationship between the public and industry: 1) The right to know about the health effects and environmental impact of materials used by the plant under investigation; 2) The right to inspect facilities where citizens and their experts actually have access to view, test, and witness the operation or process of the manufacturing site; and 3) The right to negotiate agreements with responsible parties over issues affecting human and environmental health, safety, and welfare.[12]

These guarantees are aimed not at substituting for government action but rather at establishing the community's right to defend itself against chemical trespass. The local community cannot always wait for city councils, state legislatures, or Congress to act. Their motives are often compromised by complicity with the polluters, and their regulatory agencies are uncoordinated and occasionally in conflict with each other.

Since the early 1980s there have been scores, if not hundreds, of examples where citizens won concrete agreements from industry to provide a level of environmental, health, or safety protection well beyond federal or state requirement.

- *Lewis Chemical Corporation*, Hyde Park, Massachusetts. On March 27, 1981, the Hyde Park Chapter of Massachusetts Fair Share, a statewide citizen group, completed the first inspection in the nation where a community delegation, along with an industrial hygienist-environmental engineer, performed a wall-to-wall inspection. The inspection found preventable worker-safety and environmental threats. The company refused to investigate the chemical storage incompatibly located next to bare live wires, and the plant exploded and burned on May 25, 1983.
- *Chevron Corporation*, Richmond, California. In the spring of 1986, The West County Toxics Coalition, a predominantly African American community organization, completed a one-day inspection of this giant refinery. The community delegation, including three technical experts and ten community residents, formulated fifty-three demands for improvements at the facility. When several sets of negotiations failed to result in an agreement, the community group persuaded the regional Air Quality Board to force Chevron to install vapor recovery equipment on its loading docks to prevent the release of benzene, a known carcinogen, into the air.
- *Sheldahl, Inc.*, Northfield, Minnesota. Between the mid-1980s and 1992, the Amalgamated Clothing and Textile Workers' Union, Local 1481, along with a concerned citizen group, CAN (Clean Air in Northfield) won a written agreement to prevent pollution at the plant. As of January, 1992, the company, because of the changes required in the agreement, had eliminated 75 percent of their 1989 methylene chloride usage and emissions levels.

Two advantages derive from neighborhood campaigns to change sources of pollution by asserting new rights. First, exercising the right to know, inspect, and negotiate encourages the prevention of chemical hazards before they occur. The expensive and hazardous post-mortem approach adopted by government can be avoided entirely by a vigilant and organized community that focuses on preventing the pollution and workplace exposures before they occur.

A second advantage of rights-based campaigns is that activism generally revitalizes a community, restoring its faith in the democratic process. Those who have come through local environmental campaigns have received an education in law, science, politics, and human relations and emerge better equipped to assist their community in all its future democratic development. The whole nation benefits from public participation.

CONCLUSION

It is easy to view the present environmental crisis as incorrigible, the miserable offspring of a failed democracy and a reckless economy. Many critics have suggested that injustice and impotence so permeate our society that nothing short of an environmental Armageddon seems likely to cleanse it. Whether it is possible to have a community right to decide without corresponding rights to economic resources is not clear. Obviously, a more direct system of democracy will make great demands on people's time and resources. It will therefore require basic income or job security and greater equalization of resources such as wealth and access to media and education. In an age of great capital mobility, international strategies will also be needed. However, America's past contains the seeds for its regeneration, so that its present—though soured by the overconcentration of wealth and the failure of its political institutions—reveals vital signs of new life. Its future is as fruitful as we choose to make it. Democracy, the force that drove this country in its youth, is the power we now can harness to achieve a wise maturity.

NOTES

1. Roderick Nash, *Rights of Nature* (Madison, Wis.: Univ. of Wisconsin Press, 1989), 11-12.
2. According to A. E. Dick Howard, "[Locke's] theories of equality in a state of nature are rooted in American examples. In the 1693 'Letter to the Free Society of Traders,' Locke quoted Gabriel Sagard's *Histoire Du Canada* (1636) in reference to the Huron noncoercive elective system of government as well as William Penn's account of the Delaware Indians." See *The U.S. Constitution: Roots, Rights, and Responsibilities* (Washington, D.C.: Smithsonian Institution Press, 1992), 117.
3. Bruce Johnson, *Forgotten Founders* (Spruce, Mass.: Gambit, 1982).
4. In Frank Bryan and John McClaughry, eds., *The Vermont Papers* (Post Mills, Vt.: Chelsea Green Publishing, 1989), 3.
5. Richard K. Matthews, *The Radical Politics of Thomas Jefferson* (University of Kansas Press, 1984), 83.
6. William Ryan, Susan Birmingham, and Marc Osten, "Why $16 Million Dollars is a Small Price to Pay to Defeat the Environmental Protection Act," *Vote Environment*, 3 November 1990 (a study by the California Public Reseach Group).
7. William Greider, *Who Will Tell the People?* (New York: Simon & Schuster, 1992).
8. Greider notes that think tanks and foundations accomplish the "research and advocacy functions" normally performed by political parties. "The economic function of political parties and secondary mediating institutions is that, by performing expensive tasks for others, they spread the costs of political participation among many people...reducing entry costs." Ibid.,52.
9. Barry Commoner, *Making Peace With the Planet* (New York: Pantheon, 1991), 26.
10. Robert A. Dahl, *A Preface to Economic Democracy* (Berkeley, Calif.: University of California Press, 1985), 82-83.
11. "Right of life," a Constitutional guarantee and prominently mentioned in the Declaration of Independence, is distinct from "Right to life," the slogan for the antiabortion movement. Without taking a position on this latter issue, it should be noted that abortion was not only legal but prevalent during the early days of the United States.
12. John O'Connor and Gary Cohen, *Fighting Toxics* (Washington, D.C.: Island Press, 1990), 52.

5

Creating a Culture of Destruction

Gender, Militarism, and the Environment

Joni Seager

HABITAT EARTH IS in trouble. In the 1990s, as the media, advertisers and even industry take up the green cause, this unsettling assessment scarcely escapes the attention of many people. Even schoolchildren can rattle off a litany of environmental horrors: ozone depletion; acid rain; pollution of groundwater; the startling and escalating rate of loss of bird, animal, and plant species; tropical deforestation; massive and deadly chemical, oil, and toxic spills; and the list goes on. How do we make sense of this sort of information, and how do we think about these problems in the context of a growing consciousness about environmental justice? To start, I suggest, we need to shift our frame of reference.

My training as a geographer, in both the physical and social sciences, led me to conceptualize environmental problems in their physical forms—as problems of physical systems under stress. This is not an unusual conceptualization. Popular media reporting on environmental issues reinforces this interpretation. The media conveys powerful and enduring images: of oil-slicked otters dying in Prince Edward Sound, of acres of scorched stumps of tropical rain forests in Brazil, of NASA computer-simulated photos of the pulsing hole in the ozone layer. These images, and the reporting that accompanies them, encourage us to think of environmental problems in their physical forms: too big a hole in the sky, too many fossil fuel pollutants, too much garbage, not enough water, not enough fuelwood.

The tendency to conceptualize environmental problems in their physical form has a number of implications. Popular reporting on environmental affairs that focuses on the horrific extent of environmental calamity ignores *causality.* The journalistic predilection for using passive language wherever possible leaves us all ill-equipped to make sense of the environmental crisis. Statements such as "we will 'lose' 50,000 animal species annually by the end of the century" or "there are more toxins in the Great Lakes than ever before" or "the ozone hole is growing" may provoke an awareness of environmental problems, but they leave the source or cause unexamined. Questions about agency—that is, the social and economic processes by which we arrive at a state of scorched trees and dead otters—are placed in the distance, if they are raised at all. Neither specialist training nor popular media coverage encourage us to ask questions about the actors, institutions, and processes behind the oil-slicked beaches or the toxic substances in our communities.

But the environmental crisis is *not* just a crisis of physical ecosystems. The real story of the environmental crisis is one of power and profit and the institutional and bureaucratic arrangements and the cultural conventions that *create* conditions of environmental destruction. Toxic wastes and oil spills and dying forests, presented in the daily news as the entire environmental story, are symptoms of social arrangements, and especially of social *derangements.* The environmental crisis, more than the sum of ozone depletion, global warming, and overconsumption, is a crisis of the dominant ideology.[1] This basic truth must be prominent for those of us who bring a feminist, race, and class-conscious analysis to environmental issues.

We have so degraded our environment, so stressed physical carrying capacities that the lives of millions of people on the planet (and certainly our cherished ways of life) are at risk. But, "we"—an undifferentiated humanity—have not done so. Rather, large-scale environmental problems are the result of control exercised by people within very particular clusters of powerful institutions, that include, prominently, militaries, multinationals, and governments, which often act in collusion.

Governments, militaries, and multinational institutions are primary agents of environmental degradation, while eco-establishment institutions set the environmental agenda and frame the ways in which we perceive environmental crisis. All of these institutions share a distinctive sociological profile: They are run by white men in the United States and Europe, and by men from the class and racial elites within other countries. A radical analysis of the state of environmental affairs requires understanding that this class, race, and gender profile is neither coincidental nor inconsequential.

As a brief illustration of the ways in which it matters that these institutions are run by male elites, consider the role of militaries as players within the environmental arena. Militarized environmental destruction is more global, more ubiquitous, and more protected than the actions of even the most flagrantly irresponsible multinational corporations or governments. Whether at peace or at war, militaries are the biggest threat to the environmental welfare of the planet. The extent of militarized environmental destruction ranges from the poisoning of hundreds of local communities in the United States and the former Soviet Union by toxic military bases, to the wartime destruction of whole ecosystems in Afghanistan, Central America, the South Pacific, Vietnam, and the Persian Gulf, among others. The most startling fact about militarized environmental destruction is its universality: There is both a sameness and a predictability about military culture around the globe, whether they serve socialist, capitalist, or monarchist states. A common denominator shapes the culture of militaries, one that transcends political ideology or national allegiances. To many feminists that denominator seems to be the hegemony of privileged masculinity. Militaries consider environmental safety a low priority and remain uninterested in mundane matters like environmental conservation.[2]

All militaries use national security as an excuse for their activities and as a cloak of secrecy. The national-security knot is a perfect defense: It allows the military to deflect most questions about what they are doing that is supposedly in the interest of national security. National security, a vague and constantly shifting concept, has no real or absolute meaning; the military defines it with the agreement of other men in the national security loop. But one consistent hallmark of national security is that men define it and defend it.[3] Particulary masculinist values underlie the ethos of military security, an ethos inherently at odds with an environmental stance. As one environmentalist notes,

> while military security rests firmly on the competitive strength of individual countries at the expense of other nations, environmental security cannot be achieved unilaterally; it both requires and nurtures more stable and cooperative relations among nations.[4]

Secrecy is the gatekeeper of power. The buffer of secrecy that governments and militaries erect to protect their joint interests is inherently antidemocratic. All elites, and male elites in particular, use secrecy to privatize access to knowledge.

The military tradition of secrecy, not just created to protect male power, is integrally rooted in the cult of masculinity that produced the first nuclear weapons. Nuclear weaponry wed science with militarism for the first time. Science is the preserve of men, who have long used secrecy, mystery, and obfuscation to protect science as a wholly male preserve.[5] Physicists managed to harden the gender barriers of their discipline by allying themselves with the military; previously caricatured as woolly headed, mild-mannered sissies, physicists improved their public image inestimably by producing monster weapons in the 1940s and 1950s.[6] By combining forces, male scientists and military men forged an impenetrable alliance: Military men got the ultimate weapon, and physicists got to prove their manhood. The hypermachismo and hypersecrecy that characterize nuclear weapons programs today—behind which egregious environmental violations are hidden—is a product of the fusion of masculinized science with militarized masculinity.

In *Fathering the Unthinkable*, Brian Easlea, a former nuclear physicist, comments that "modern science is unique in its repertoire of aggressive sexual and birth imagery. Our whole culture is basically masculine in character but modern science is the cutting edge."[7] He suggests that

> physicists were not only anxious to stamp and be the first to stamp their masculine power of intellect upon the world through their historic building of the atomic bomb but that, driving them relentlessly forward, was a subterranean desire to demonstrate once and for all the unique creativity of the male vis-à-vis the female.

Carol Cohn adds to Easlea's analysis in her study of the defense community—a community she describes as burdened by "the ubiquitous weight of [male] gender, both in social relations and language," that thrives on sexually explicit and transparent imagery, and sanitized abstraction.[8] The overwhelming representation of the early years of the atomic project is that of nuclear scientists giving birth to male progeny with the ultimate power of violent domination over female Nature. Cohn talks about the phallic imagery that pervades all military discussions of military strength, a language of soft lay-downs, deep penetration, and hard missiles. The development of the atomic bomb is rife with "male birth" imagery: The Los Alamos bomb was referred to as "Oppenheimer's baby," or "Teller's baby" at the Lawrence Livermore labs; those who wanted to disparage Teller's contribution claimed that he was not the bomb's father, but the mother (i.e., that Teller merely "carried" someone else's idea). In the early tests, before they were certain that the bombs would work, the scientists expressed their concern by saying that they hoped the baby was a boy, not a girl (i.e., a dud). Cohn reports that mothering (denigrated) vs fathering imagery, male (positive) vs. female (negative) sex-typing is still entrenched in the nuclear mentality. Military nuclear weapons programs, perhaps more so than other military projects, are still very much fueled by a cult of masculinity; it is *this* institutionalized context that undergirds the tradition of secrecy and self-importance that underlies military environmental unresponsiveness.

However, it is not just *intangible* bonds of fraternity that cement relationships between men in government and men in the military. Concrete partnerships exist among the leaders of governments, militaries, and industry—the network of overlapping financial, industrial, and policy agendas serving the interests of elites that Dwight Eisenhower identified as the "military-industrial complex." Revolving-door job connections and business networks among defense contractors, military officials, and ex-government officials ensure a mutual self-interest in keeping military programs and defense contracts going and in keeping critical scrutiny of military actions and priorities to a minimum.

The collusion of military, governmental, and industrial interests creates an impenetrable knot of elite power. Antimilitarist activists over the last two or three decades have been especially effective in exposing its social and political ramifications. There has been much less effort directed at exposing the *environmental* consequences, and yet it is increasingly pressing that we do so.

The most immediately apparent environmental consequence of the seamless web of overlapping power is that militaries are left alone to be their own environmental watchdogs. No country explicitly exempts military facilities from environmental regulations and oversight. The universality of military environmental *immunity* makes a mockery of distinctions between "democratic" and "authoritarian" regimes. In the United States, hundreds of towns have been poisoned by the military; places such as Fernald, Ohio; Hanford, Washington; or Rocky Flats, Colorado, have become symbols not only of environmental catastrophe but of the flagrant abuses that derive from a hands-off military environmental policy. The environmental impact of unimpeded military might in the former Soviet Union is even more devastating.

Everywhere in the world, the intertwined interests of men in government, military, and industry serve to improve their wealth and social standing. Specifically, many of the men in the "old boy networks" make a lot of money from industries and resource-extraction activities that cause massive environmental degradation. One of the policy ramifications of these intertwined networks is to discourage environmental regulation or enforcement of existing environmental laws. For example, the ferociously rapid exploitation of the rainforests in Malaysia, Thailand, and Myannar (formerly Burma) can be attributed to a powerful network of interlocking government-military- industrial interests operating in each of those countries. Central America's teetering on the brink of environmental collapse is also closely related to the entrenched networks of militarized industrial and governmental power that operate throughout that region. Environmental problems are directly rooted in a long history of natural resource plundering by foreign corporations operating in conjunction with local elites—and a concomitant history of the use of military force to prop up those elites and to protect foreign economic interests. In addition, as one of the most heavily militarized regions in the world, the ecology of Nicaragua, El Salvador, Honduras, and Panama have experienced massive environmental war damage.

Beyond observing the working of old boy networks, the creation of a *masculinized* knot of power results in a special type of environmental conflict. Worldwide, most community and grass-roots environmental activists are women; worldwide, all militaries are institutions of men, even in those countries where a few women participate in the armed forces. Militaries do not like to be challenged, especially by women. The dynamic of powerful men protecting secretive and intertwined vested interests from outsiders is heightened when the community activists who challenge military power are women (or men) of a different class or race.

Populations excluded from the inner circles of power disproportionately bear the impacts of militarized environmental destruction, perhaps the most pernicious environmental effect of the lock on power held by militarized male elites. Everywhere in the world social justice and environmental protection are inseparable. This is true in the Amazon, where a wealthy landowning elite supported by the military is trying to suppress indigenous movements for sustainable use of rain forests, and in Central America, where a wealthy militarized landowning elite has drenched the land in pesticides and herbicides, trying to squeeze every drop of cash-crop profitability out of a sometimes marginal land base. It is no less true in London or Los Angeles, where marginalized people confront environmental degradation on a daily and domestic level.

Racial minorities are disproportionately victimized by environmental health hazards. In the United States, blood lead levels in minority children are dramatically higher than in the population as a whole, as are certain cancers and respiratory diseases. Race is one of the most significant variables in determining the location of commercial, industrial, and military hazardous-waste sites. In the United States, the three sites accounting for more than 40 percent of the nation's total disposal capacity are located in predominantly African American or Latino American communities; the nation's largest hazardous-waste landfill, in Sumter County, Alabama, is in the heart of the state's "black belt"; the predominantly African American and Latino American Southeast side of Chicago has the greatest concentration of hazardous-waste sites in the nation.

Since 1963, the United States has exploded 651 nuclear weapons or devices on the U.S. mainland; all the nuclear-bomb testing sites in the continental United States are located on Native American lands, mostly on western Shoshone territory in Nevada.[9]

Environmental racism extends beyond national borders. The practice of dumping toxic wastes from Europe and the United States in Third-World countries perpetuates old patterns of imperialism, colonialism, and racism.

Islanders in the South Pacific have borne the greatest brunt of British, French, and U.S. nuclear testing programs. Most of the nuclear-weapons systems stationed in Europe and in the United States have been tested on indigenous peoples' lands in the Pacific, without their consent and often without warning. Over the past four decades of nuclear testing, some Pacific islands have disappeared entirely, and many more have been rendered uninhabitable, and they will remain so for tens of thousands of years. Entire island populations have been dislocated, social disruption in the South Pacific as a result of military activities has been enormous, and the health consequences are incalculable. As a general rule, weapons testing, chemical dumping, or war-simulation exercises that would never be tolerated in places like southeast England or the northeast United States, are simply exported, often in a close replication of colonial trading patterns. Environmental concerns are not taken seriously when the environment of concern belongs to someone else, especially small developing countries.

Women's social roles as family caretakers and, in agrarian economies, as primary subsistence providers, situate them in the environmental front lines. In the aftermath of the Persian Gulf war, for example, women in both Kuwait and Iraq, even urban women unaccustomed to fashioning family provisions from raw materials—became the hewers of wood and carriers of water. A brief United Nations' report from Iraq, several months after the war, observed that women and children were spending large parts of their day searching out food, fuel, and water, often carrying these supplies for miles. Indigenous peoples, too, paid a particular price of militarized environmental damage. The lands of the Bedouins were devastated: the desert ecosystem on which they rely was mined, bombed, and debased by the military presence in Saudi Arabia and the war in Kuwait and Iraq. Military activity, pollution, and oil spills dislocated the Marsh Arabs, a little-known population that lives in the wetlands and marshes of the Persian Gulf.

Many of the toxins used routinely by militaries have particular health effects on women. For example, Vietnamese women suffer enormous health crises as a result of the wartime poisoning of Vietnam. The U.S. military dumped approximately twenty-five million gallons of defoliants, herbicides, and assorted noxious chemicals on Vietnam over the course of the war, among the most toxic of which was "Agent Orange." Dioxin, a primary contaminant in Agent Orange, persists in the food chain for decades. It is highly carcinogenic, even in minute quantities, causing genetic mutations and any number of cancers. Dangerous levels of dioxin are passed from generation to generation. But dioxin is also a major teratogenic (birth-deforming) chemical. Vietnamese women today have the highest rate of spontaneous abortions in the world; birth defects occur at alarming rates; 70 to 80 percent of women in Vietnam suffer from vaginal infection; cervical cancer rates are among the highest in the world; and fetal death rates in pregnancy were forty times higher in the early 1980s than they were in the 1950s. This pattern is not particular to Vietnam.

Reproductive failures of all kinds are often the first indicators of severe environmental trauma. These health problems also have wider social consequences. In cultures where women are valued primarily for their reproductive capacities (and where is this not the case?), when those capacities are diminished, women who cannot or will not bear more children may face ostracization and a loss of their social stature. At a minimum, women face increased pressure to bear more children. In the aftermath of the Bhopal industrial disaster, for example, Indian feminists reported increasing rates of domestic violence, divorce, and abandonment suffered by poor women who were maimed by the chemical explosion, a side effect of the mounting frustrations of increased poverty and ill health.[10]

Militaries consume social resources at an astonishing rate, and, worldwide, the military share of national budgets is increasing. Ruth Sivard, a tireless researcher who monitors international military spending, estimates that in constant dollars world military expenditures in 1987 were about 2.5 times the level of 1960.[11] There are direct environmental trade-offs of this level of spending, since military and civilian programs have to compete for shares of limited national revenues. A few comparisons make clear the environmental cost of military spending: For the price of one British Aerospace Hawk aircraft, 1.5 million people in the Third World could have clean water for life; the money spent on one nuclear weapons test could provide installation of 80,000 hand pumps to give Third World villages access to safe water; the money spent on operating a B-1B bomber for one hour could provide maternal health care in ten African villages to reduce infant mortality by half; the money needed to supply contraceptive materials to women around the world already motivated to use family planning is the equivalent of ten hours of global military spending; in ten days of the

Persian Gulf war, the U.S. military spent the equivalent of the entire annual budget designated for energy development and conservation. In India, a country with over one-third of its population living below the poverty line, the government spends 14 percent of its revenue on defense; in Saudi Arabia military spending, as a percent of the gross national product (GNP), jumped from 5 percent in 1960 to 22 percent by the late 1980s; and the Canadian government, usually considered to be a bit player in global militarism, spends twelve times as much on the military as it does on the environment.[12]

Escalating military spending and increasing military control of national budgets undermines the ability of citizens, elected government officials, and bureaucrats alike to monitor and control military activities. Moreover, the price of military expansion everywhere in the world is environmental neglect, increasing social inequality, and deterioration in the daily quality of life for hundreds of thousands of people. Ironically, as militaries gain economic clout, and their environmental record deteriorates, environmental monitoring programs themselves, usually chronically underfunded, are often early victims of budget cuts on the social-spending side of the national ledger. The effects of reductions in social welfare programs, and deterioration in the environmental quality of life, ripple through society unevenly: People on the economic margins, the poor, and the disenfranchised bear the brunt of social dislocation. Since women and minorities worldwide comprise the largest population living in poverty, when spending on social programs is reduced as a trade-off for military spending, it is they who suffer first—and last.

Militaries, multinationals, and governments are on an unrestrained global wilding spree, one that cannot be halted by a few new law-and-order regulations. Environmental regulations are necessary, but not sufficient, precisely because they leave intact the culture of institutions of destruction. Strategies for *real* environmental protection have to be rooted in understanding how race, gender, and class privilege are integral to the functioning of these institutions. If this seems to be a daunting agenda, it is primarily because our understanding of causality in assessing the environmental crisis has been so inadequate. As progressive environmentalists, we must learn to be more curious about causality and about agency. Grass-roots environmental groups have been the most effective at naming names but perhaps the least effective at exposing the larger linkages, the structures, and the *culture* behind the agents of environmental destruction. Those structures and cultures are not gender or race neutral. In action-oriented environmental groups, it is difficult to ask questions about relationships between men and women, whites and African Americans, and our environmental crisis. It is not easy to ask about masculinity and militarism, privilege and power, and it is not considered polite to point out that white men (or, outside the European sphere, men from other ethnic and racial groups) have been far more implicated in the history of environmental destruction than women. But there is too much at stake to stick to the easy questions and polite conversation.

NOTES

1. Pam Simmons, "The Challenge of Feminism," *The Ecologist* 22(1) (January/February 1992): 2–3.
2. An observation made by Stephanie Pollack and Seth Shulman in "Pollution and the Pentagon," *Science for the People* 19(3) (May/June 1987): 5–12.
3. See Barbara Ann Scott, "Help Wanted: Women Defense Experts and Decision-makers," *Minerva* 6 (Winter 1988).
4. Michael Renner, quoted by Philip Shabecoff, "Environment," *New York Times*, 29 May 1989: 6.
5. For compelling analyses of the masculine stranglehold on science, see Evelyn Fox Keller, *Reflections on Gender & Science* (New Haven: Yale Univ. Press, 1984); and Sandra Harding, *The Science Question in Feminism* (Ithaca, N.Y.: Cornell Univ. Press, 1986).
6. A point suggested by Lisa Greber in, "The Unholy Trinity: Physics, Gender, and the Military" (B.A. honors thesis, Massachusetts Institute of Technology, 1987).
7. Brian Easlea, *Fathering the Unthinkable: Masculinity, Scientists and the Nuclear Arms Race* (London: Pluto Press, 1983).
8. Carol Cohn, "A Feminist Spy in the House of Death," in *Women and the Military System*, ed. Eva Isaksson (New York: Harvester, 1988). See Cohn's bibliography in this article for a good list of feminist critiques of Western male rationality in the military system.
9. Bernard Nietschmann and William Le Bon, "Nuclear Weapons States and Fourth World Nations," *Cultural Survival Quarterly* 11 (1987): 5–7.
10. Padma Prakash, "Neglect of Women's Health Issues," *Delhi Economic and Political Weekly*, 14 December 1985.
11. Ruth Sivard, *World Military and Social Expenditures, 1987-88* (Washington, D.C.: World Priorities Institute, 1987).
12. Figures from Ruth Leger Sivard, *World Military and Social Expenditures, 1991* (Washington, D.C.: World Priorities, 1991); K. S. Jayaraman, "Poor and Buying Weapons," *Panoscope* 14 (September 1989); Jim Hollingworth, "Global militarism and the Environment," *Probe Post* (Fall 1990): 42; "The Cost of the Persian Gulf War," Boston, Massachusetts SANE/FREEZE fact sheet, 1991.

6

Environmental Consequences of Urban Growth and Blight

Cynthia Hamilton

THE LINK BETWEEN economic development and environmental degradation is intensifying debate between those in the United States who support continued urban growth and citizens seeking to protect their communities from its uncontrolled consequences. This debate has sharpened recognition that growth and development are simultaneously sources of wealth and improved living conditions, as well as death, destruction, and inequality, whether the subject is forests, water and air, or neighborhoods. As Hazel Henderson explains, effective policy is never a matter of growth vs. no growth but rather of what is growing, what is declining, and what must be maintained.[1] For people of color, little has been maintained, as communities and neighborhoods decline.

The most advanced stage of industrialization has been the most dangerous in communities of color, resulting in severe deindustrialization as well as extraordinary contamination from industrialization. Three out of five of the largest commercial hazardous landfills in the United States are located in predominantly African American or Latino American communities; these landfills account for 40 percent of the total estimated commercial landfill capacity in the nation. Three out of every five African Americans and Latino Americans live in communities with uncontrolled toxic wastes.[2] In a study conducted in 1990, "scientists performed autopsies on 100 youths between the ages of 15 and 25 who had died as a result of violence, accident, and other non-medical causes.... Eighty percent had 'notable lung abnormalities,' and 27 percent had 'severe lesions on their lungs....' The air in Los Angeles is slowly killing us."[3] In Los Angeles, regulatory agencies estimate that African Americans and Latino Americans have the greatest pollution exposure: They live nearest the polluting industries and constitute a significant percentage of the work force in industries where exposure to toxics is greatest.[4]

This essay explores the contradictions of growth and environmental destruction as manifested in the U.S.'s urban areas. In particular, it addresses the crisis of the inner cities that many African Americans called home until they were displaced by new growth, redevelopment, and renewal. The inner city, once the site of communities of the poor, have been reclaimed and reshaped for office towers, financial districts, and condominiums. The old residents have been replaced by more affluent, young singles or couples without children. The land has become more profitable with its new use. Old neighborhoods have been lost. To produce greater value (i.e.,tax dollars and other revenue), the growth process makes people and places expendable, reduced to the value of commodities in a consumer society.

These issues are not normally associated with discussion of the environment. However, as the *Gaia Atlas of Green Economics* makes clear, ecological concerns cannot be separated from ethics or the politics of development. According to the atlas,

> out of the industrial economy issues the dazzling variety of consumer goods that make for the affluent society. To produce them, it sucks up...life on Earth and indeed, the Earth itself. It lays nature waste, exploits women, undermines family and community, and impoverishes hundreds of millions of people. It is socially, culturally, environmentally, and economically unsustainable.[5]

In many cities, growth and development exist beside decay and blighted slums. The two are part of the same process. New development, causing displacement and relocation, may be the source of employment for some while eliminating employment permanently for others. The way in which this displacement and inequity occurs is deeply related to injustices long connected with racism and class.

Most of our discussions and debates about urban growth are project-specific proposals for office towers, condominimum conversion, shopping centers or mini-malls, land fills, and waste incinerators. A project's controversy often shapes our framing of the issues: a "good" or "bad" response depends on the projects consequences—displacement, increased traffic, or higher rents.

It is important, however, to examine issues of urban growth within the broader context of U.S. industrial policy. The imperative to grow or to perish, which dictated the early industrial revolution, is no less predominant today. The economic survival of the United States rests on this mandate for growth, which has had severe human consequences: environmental destruction, health problems, declining quality of life, and increases in the cost of living. The more we grow, the more resources we use and abuse, the greater the cost of survival. Of course, growth has advantages, but we must begin to ask if growth can occur along with equality rather than greater inequalities. If we cannot grow and develop with constructive rather than destructive production, we must begin to consider the alternatives to industrial society as we know it. But the idea of growth has been promoted as ideology as well as policy. Growth has almost become synonymous with progress, and critics are summarily dismissed as irrational opponents of progress.

The consequences of endless growth have resulted in colonialism, both domestic and international, because local markets become saturated and local resources depleted. While people of color and the Third World generally bore few of the fruits of development, they are now asked to be partners in solutions to an environmental crisis created by others. Communities of color in the United States house a disproportionate share of toxic waste. And exporting industrial residues, toxic and radioactive waste to the Third World has become a new form of imperialism. This, unfortunately, characterizes the relationship to growth and development that most people of color experience. Whether manifested as the uncontrolled urban sprawl of U.S. cities or the greed that has turned rain forests into cattle ranches, development has come to people of color at great cost. For some, progress means scarcity of affordable housing, land, clean water and air; for others, it means serving as the repository of waste and dangerous by-products. Progress comes often at the expense of workers' health and safety. Their exposure to toxic substances is twofold: on the job and in the communities surrounding their place of employment. Therefore, communities of color experience the irony of destruction coupled with industrial growth—what I refer to as environmental racism.

A recognition of this particular manifestation of racism is what brings people of color to organize efforts to preserve their communities, to defend their homes and their lives against displacement, destruction, and disease. It is not a defense of space in the abstract; environmentalism among people of color manifests itself as a specific response to uneven development, to the lack of equity in relation to resources that we presume to hold in common (e.g. air and water) but that in practice are treated as private commodities. The privatization of the use of public space in cities is an example. The wholesale destruction of communities, either by neglect, displacement, or redevelopment, is most often the source of activism. In either case, the place that had been home for some is appropriated for private use. Inner-city residents have been overburdened with the residue, debris, and decay of the industrial revolution. They bear the burden of growth, soldiering America's waste and its abandoned factories and warehouses. For example, an article in *The Chicago Sun Times* (31 May 1987) described residents of the city's Southside as "victims of environmental neglect that has made the far Southside a mine field of toxic hazards which include abandoned factories, toxic waste dumps, industrial air pollution and tainted water."

The common experience of the costs of growth and development for people of color in the United States has produced the basis for new alliances and collaboration based on a new level of political analysis. The old civil-rights consciousness that shaped social movements in the African American community has been replaced by a consciousness of corporate power and economic inequalities. There is a new recognition of the negation of political rights by corporate power and the global nature of corporate power, which influences the development of new social movements.

In Los Angeles, workers and community residents are coming together to challenge corporations that are the major air polluters. The regulatory agencies have estimated that African American and Latino Americans have the greatest pollution exposure: They live nearest the polluting industries and constitute a significant percentage of the work force in industries where exposure to toxics is greatest. According to a study by the Labor/Community Strategy Center, "more than 81% of all the workers in the state [of California] who are exposed to lead are employed in L.A. County where low wage labor and irresponsible management form a lethal combination."[6]

In 1989, hearings were conducted by the SouthWest Organizing Project in Albuquerque on toxics in minority communities. Testimony from workers and community residents revealed the soaring damage resulting from military toxics. Contaminated ground water, unlined beds of toxic sludge, and carelessly constructed landfills will cost millions in cleanup costs, but, for many families, the damage has been done.

Conditions in urban communities of color express the real critique of industrialization and growth because the contradictions are so stark. Here we *feel* the irony of living in a country that represents 5 percent of the world's population and whose residents consume 30 percent of its resources and 60 percent of its energy resources.

The ideological meshing of growth and progress has made economic blackmail of workers and community residents an easy tool. Corporations ask communities of color to make a contribution to the common good by agreeing to house nuclear and hazardous waste as well as waste disposal plants; workers are told that the cost of clean industry may be plant closures.

It is no accident that the crisis of industrialization manifests itself first in communities of color. These communities have experienced the most severe deindustrialization and the greatest contamination from industrialization. And the consequences of development models emphasizing capital intensive projects (e.g., highways, office towers, shopping malls and condominiums) have made grass-roots organizations in communities of color wary of accepting the benefits of such development.

According to the 1987 report of the Commission for Racial Justice of the United Church of Christ:

- More than fifteen million African Americans live in communities with one or more uncontrolled toxic waste sites.
- More than eight million Latino Americans live in communities with one or more uncontrolled toxic-waste sites.
- African Americans are heavily overrepresented in the populations of metropolitan areas with the largest number of uncontrolled toxic-waste sites: Memphis, Tennessee (173 sites); St. Louis, Missouri (160 sites); Houston, Texas (152 sites); Cleveland, Ohio (106 sites); Chicago, Illinois (103 sites); and Atlanta, Georgia (94 sites).

- Los Angeles has more Latino Americans living in communities with uncontrolled toxic waste sites than any other metropolitan area in the United States.
- Approximately half of all Asian Americans, Pacific Islanders, and American Indians live in communities with uncontrolled toxic waste sites.
- Overall, more than half of the total population in the United States resides in communities with uncontrolled toxic-waste sites.[7]

These realities will continue to have catastrophic health consequences for minority communities, though most will not manifest themselves for many years. Even now, Navajo teenagers have organ cancer seventeen times the national average; 50 percent of the children suffering from lead paint poisoning, resulting in low attention spans, limited vocabulary, and behavior problems, are African American. These are the *real* implications of the United States's institutional racism and development strategies, which ignore social costs and which have thereby eroded the quality of life in America's urban areas.

These problems result from land-use decisions ensuring that those populations with the most consistent residential proximity to industry in U.S. cities are poor people of color. The central business district of the early twentieth-century city was usually the center of the industrial circumference, with people of lower socioeconomic status living closer to that center. African Americans replaced the European ethnic working class completely in this vicinity by the end of World War II. In many instances housing had been built on marshes and garbage dumps. More frequently in the Northeast and Midwest, elevated trains instead of subway tunnels traversed the African American community, with business as usual continuing beneath.

The presence of noise, dirt, railroad tracks, factory pollution, warehouses, and stockyards have become the trademark of African American communities, ensuring de facto segregation. As whites moved away from the inner core of the industrial city to escape its noise, foul air, water, and land, African Americans were allowed to move into the houses left behind. This has made the boundaries of African American communities—natural divisions such as rivers or lakes, on whose banks we are sure to find factories, warehouses, and stockyards—distinct.

Only after the organized union drive of the 1930's did industry discover the advantages of relocating to the urban periphery or finding new locations in the underdeveloped Sunbelt, taking advantage of cheap land and cheap labor and moving away from the industrial Northeast. Class relations shaped the urban landscape and the emergent conflict that characterized the new urban industrial cities. The power of workers was enhanced by their concentration in large numbers and the centralized nature of production. The historical transformation in the structure of cities has been propelled not by technological development alone but by the need to reproduce existing class relations in the capital accumulation process. To escape workers' demands, industries began to move away from the cities' urban core into the suburbs, aided by governmental subsidies for the construction of industries producing for the war effort. Public-housing and highway legislation facilitated the transit of workers and commodities. The desertion of the inner city left a decaying core. Dispersing workers inhibited their capacity to organize effectively.

The decay of the inner city has become an internationally recognized condition, as indicated by the World Commission of Environment and Development.

> [Industrial] cities account for a high share of the world's resource use, energy consumption, and environmental pollution....Many [industrial cities] face problems of deteriorating infrastructure, environmental degradation, inner-city decay, and neighborhood collapse. The unemployed, the elderly, and racial and ethnic minorities can remain trapped in a downward spiral of degradation and poverty as job opportunities and the younger and better-educated individuals leave declining neighborhoods. City or municipal governments often face a legacy of poorly designed and maintained public housing estates, mounting costs, and declining tax bases.[8]

The report also notes that the resources exist to solve the urban-environmental crisis in industrial countries and therefore "the issue for these countries, is ultimately one of political and social choice."[9] In most instances the choice has been to build a new urban core, to displace the poor and create a new corporate city to serve a new set of functions. Cities are needed not to centralize production but rather to house administrative and financial headquarters such as banks, stock exchanges, and insurance companies. The reorganization of capital and the need to maintain control of workers facilitated urban restructuring. A new corporate form replaces the old industrial city, not only in appearance but with new inhabitants, displacing the poor and working class. Decentralization and sprawl have replaced centralization. This new form has also displaced a major source of social justice that communities have historically provided for their residents. Left to their own devices, poor communities in the 1930s and 1960s organized to fill gaps in services left unprovided by the local government. Communities not only provided a sense of cultural identity for residents but also offered an antidote for alienation, loneliness, and racial harassment. They offered safety and security to residents. Historically, communities provided their own safety nets through self-help organizations. For example, they organized funds for injured workers and widows; neighbors banded together to collect money for rent or came together to resist eviction.[10] As late as the 1960s, African American communities recognized community control as a self-help strategy. The collapse of community has left residents completely dependent on government service and subject to the consequences of urban restructuring.

But the loss of community was no accident. Much like the movement of industry to the periphery of cities in the 1940s and '50s, which dispersed workers as they moved to the new suburbs following jobs, African American communities have been dispersed.[11] Suburbanization is an instrument of dispersal, but new development at the city's center has also displaced and dispersed African Americans. Without community, African Americans have lost an important center of cultural and political identity that could offer an alternative source of affirmation and resistance. After 1965 and the passage of the Voting Rights Act, urban concentration also proved to be a political asset as African Americans elected representatives in urban areas. Without community, this type of action is impossible.

Uneven development and exchange have shaped power relations between nations and individuals over the past three centuries. We have known these relations by different names: colonialism, imperialism, underdevelopment, racism, or internal colonialism. This is the result of centuries of uneven development and exchange and a century of intense industrial production. The most advanced stage of industrialization has been the most toxic, resulting from the minimally controlled petrochemicals, electronics, and aerospace industries.

Development's horrors in communities of color are rampant. Children of farm workers suffer birth defects as a consequence of their mother's pesticide exposure at work during pregnancy. In and around farm worker communities, child cancer rates are high. Children whose mothers have worked in high-tech industries, where the use of dangerous chemicals is common, have high rates of birth defects. Children living around military installations have higher rates of cancer and other illnesses.[12]

To halt this self-destructive march of industrial growth and development requires citizen action guided by a critical approach to community development and industrial production. Such action must transcend isolated, individual crises and attempt to confront the national consequences of corporate, industrial behavior. So far, environmental activists and thinkers have been slow to develop a theory of political action or community development because of the focus on instrumentalities: rules, bureaucracy, and administration (in the tradition of liberal and conservative thought.) What is needed is an economic democracy that must include, first, an end to macroeconomic approaches to planning and assessment and the institutionalization of decentralization, local and regional approaches to development. Cities have become a center of injustice as a result of current development models; alternatives should be conscious of the multiple needs at this level. New planning must be undertaken in cities incorporating neighborhood voices. Second, we must focus our attention on renewable resource methods (e.g., solar energy and recycling) over nuclear energy.

Third, our political assessment and approach within a new framework of an ecology of democracy must recognize class interests in Western and developing societies. Alternatives require us to acknowledge the political intentions and consequences of growth and development strategies, intentions that include the destruction of working-class communities in the inner city. This approach necessitates rejecting an assessment of development as simply a technological advance and thereby politically neutral.

Fourth, a new social contract is necessary between citizens and the state as well as between citizens and representative and participatory forms. Corporations have used special-interest processes to influence political officials and thereby government decisions. Consequently, present political leadership is completely unresponsive to communities. Before accountability can be assured, new leadership and new forms of participation will be necessary; multiple decision-making units, such as neighborhood councils, are needed to regulate development and ensure citizen input on growth decisions. Centralized units of political decision making as well as centralized planning methods must be replaced by decentralized units. This is the essence of what Henderson means by "thinking globally and acting locally."[13]

The challenge feared most by the corporate sector is one that substitutes the collective good for their much-promoted philosophy of individualism. When groups previously left out of formal parliamentary and electoral processes demand access or develop new methods of political action, capitalists feel threatened. Struggles of urban African Americans in cities against the consequences of unbridled growth have much to contribute to the new environmental movement. For African Americans in the legally segregated "separate but equal" society, the personal commitment to change was reflective of social concerns. Demanding broad social change was a prerequisite for expanded personal rights and freedoms. Unfortunately, industrial society forced the separation of private and social concerns. As a result, individuals are locked into selfishly considering individual rights rather than cooperating to meet community needs.

The essence of a new social contract must therefore be the reaffirmation of the common good. Livable cities will only be possible when the collective good is understood to have meaning for each individual. A community agenda must replace the current corporate agenda for American cities, so that we would be forced to consider issues of sustainability: employment, livable space, resource management that avoids excessive waste, and pollution control as a health measure. But this community agenda would also revive a notion of the collective good; social concerns would become more central than private good. Individual and collective good can no longer remain separate. The false contradiction between social good and individual rights and needs has produced the current problems and crises. This community agenda might therefore be prefaced by a new social contract that reaffirms the common good. Individual property rights must no longer be permitted to infringe on the quality of life affecting everyone.

NOTES

1. Hazel Henderson, *The Politics of the Solar Age: Alternatives to Economics* (New York: Doubleday/Anchor, 1981), 7.
2. United Church of Christ, Commission for Racial Justice, *Toxic Wastes and Race in the United States*, New York, 1987.
3. Eric Mann, *Los Angeles's Lethal Air: New Strategies for Policy, Organizing, and Action* (Labor/Community Strategy Center, 1991), 5.
4. Ibid.

5. Paul Elkins, Mayer Hillman, and Robert Hutchison, *The Gaia Atlas of Green Economics* (New York: Doubleday/Anchor, 1992), 12.
6. Mann,"Lethal Air," 28.
7. United Church of Christ, Commission for Racial Justice, *Toxic Waste and Race in the United States*, A National Report on the Racial and Socio-Economic Characteristics of Communities with Hazardous Waste Sites, New York, 1987, p.xiv.
8. World Commission on Environment and Development, *Our Common Future* (New York: Oxford Univ. Press, 1987),24.
9. Ibid., 25.
10. St. Clair Drake and Horace Cayton, *Black Metropolis: A Study of Negro Life in a Northern City* (New York: Harcourt Brace, 1962).
11. David Harvey, "The Urban Process under Capitalism: A Framework for Analysis," in *Urbanization and Urban Planning in Capitalist Society*, ed. Michael Dear and Allen Scott (New York: Metheun, 1981).
12. At the University of Southern California, scientists performed autopsies on one hundred youth between the ages of 15 and 25 who had died as a result of violence, accidents, and other nonmedical causes. What they discovered was shocking: 80 percent had "notable lung abnormalities" and 27 percent had "severe lesions on their lungs." See Russell Sherwin "Findings" (International Specialty Conference on Tropospheric Ozone and the Environment, Los Angeles, March 21, 1990).
13. Henderson, *Politics of the Solar Age*, 355.

7

Feminism and Ecology

Ynestra King

NO PART OF living nature can ignore the extreme threat to life on earth. We face a worldwide deforestation, the disappearance of hundreds of species of life and the increasing pollution of the gene pool by poisons and low-level radiation. We also face biological atrocities unique to modern life—the AIDS virus and the possibility of even more dreadful and pernicious diseases caused by genetic mutation as well as the unforseen ecological consequences of disasters such as the industrial accident in Bhopal, and the nuclear meltdown in Chernobyl. Worldwide food shortages, including episodes of mass starvation, continue to mount as prime agricultural land is used to grow cash crops to pay national debts instead of food to feed people.[1] Animals are mistreated and mutilated in horrible ways to test cosmetics, drugs, and surgical procedures.[2] The stockpiling of ever greater weapons of annihilation and the horrible imagining of new ones continues. The piece of the pie that women have only to sample as a result of the feminist movement is rotten and carcinogenic, and surely feminist theory and politics must take account of this however much we yearn for the opportunities within this society that have been denied to us. What is the point of partaking equally in a system that is killing us all?

The contemporary ecological crisis alone creates an imperative that feminists take ecology seriously, but there are other reasons ecology is central to feminist philosophy and politics. The ecological crisis is related to the systems of hatred of all that is natural and female by the white, male, Western formulators of philosophy, technology, and death inventions. The systematic denigration of working-class people and people of color, women, and animals is connected to the basic dualism that lies at the root of Western civilization: nature vs. culture. But this mind-set of hierarchy originates within human society. It has its material roots in the domination of human by human,

---

This article has been adapted from previous writings produced over the last ten years.

particularly women by men. While I cannot speak for the liberation struggles of people of color, I believe that the goals of feminism, ecology, and movements against racism and for the survival of indigenous peoples are internally related, and must be understood and pursued together in a worldwide, genuinely pro-life[3] movement.

Around the world capitalism, the preeminent culture, and economics of self-interest are homogenizing cultures and simplifying life on earth by disrupting naturally complex balances within the ecosystem. Capitalism is dependent upon expanding markets, therefore ever-greater areas of life must be mediated by sold products. From a capitalist standpoint, the more things that can be bought and sold, the better. So capitalism requires a rationalized worldview that asserts that a human science and technology are inherently progressive, which systematically denigrates ancestral cultures and asserts that human beings are entitled to dominion over non-nature.

Nonhuman nature is being rapidly simplified, undoing the work of organic evolution. Hundreds of species of life disappear forever each year and at an accelerating rate. Diverse complex ecosystems are more stable than simple ones. They have had longer periods of evolution and are necessary to support human beings and many other species. Yet in the name of civilization, nature has been desecrated in a process of rationalization sociologist Max Weber called "the disenchantment of the world."

The diversity of human life on the planet is also being undermined. This worldwide process of simplification impoverishes all of humanity. The cultural diversity of human societies around the world developed over thousands of years, and is part of the general evolution of life on the planet. The homogenizing of culture turns the world into a giant factory and facilitates authoritarian government. In the name of helping people, the industrial countries export models of development that assume that the American way of life is the best way of life for everyone. In the United States, MacDonald's and shopping malls cater to a uniform clientele that is becoming more uniform all the time. "To go malling" has become a verb in American English—shopping has become our national pastime, as prosperous U.S. consumers seek to scratch an itch that can never be satisfied by commodities.[4]

A critical analysis of and opposition to the uniformity of technological industrial culture—capitalist and socialist—is crucial to feminism, ecology, and the struggles of indigenous peoples. There is no way to unravel the matrix of oppressions within human society without at the same time liberating nature and reconciling that part of nature that is human with that part that is not. Socialists do not have the answer to these problems; they share the antinaturalism and basic dualism of capitalism. The technological means of production used by capitalist and socialist states, although developed by capitalism, is largely the same. All hitherto existing philosophies of liberation, with the possible exception of some forms of social anarchism, accept the anthropocentric notion that humanity should dominate nature, and that the increasing domination of nonhuman nature is a precondition for true human freedom.[5] No socialist revolution has ever fundamentally challenged the basic prototype for nature vs. culture dualism: the domination of men over women.

This old version of socialism has apparently ended but still argues that the old socialist spirit of history, a valuable legacy, is not dead. It has passed onto new subjects—feminists, greens, and other bearers of identity politics, including movements against racism and for national liberation and the survival of indigenous peoples. And in this sense, these most antimodern of movements are not modern, not postmodern. In response to the modern crisis, they argue for more heart, taking the side of Pascal against Descartes, "The heart hath its reasons which the reason knows not."

ECOFEMINISM: ON THE NECESSITY OF HISTORY AND MYSTERY

Women have been culture's sacrifice to nature. The practice of human sacrifice to outsmart or appease a feared nature is ancient. And it is in resistance to this sacrificial mentality—on the part of both the sacrificer and sacrificed—that some feminists have argued against the association of women with nature, emphasizing the social dimension to traditional women's lives. Women's activities have been represented as nonsocial, as natural. Part of the work of feminism has been asserting that the activities of women, believed to be more natural, are in fact absolutely social. This process of looking at women's activities has led to a greater valuing of women's social contribution, and is part of the antisacrificial current of feminism. Giving birth is natural, although how it is done is very social, but mothering is an absolutely social activity.[6] In bringing up their children, mothers face ethical and moral choices as complex as those considered by professional politicians and ethicists. In the wake of feminism, women will continue to do these things, but the problem of connecting humanity to nature will have to be acknowledged and solved in a different way. In our mythology of complementarity, men and women have led vicarious lives, where women had feelings and led instinctual lives, and men engaged in the projects illuminated by reason. Feminism has exposed the extent to which it was all a lie—that is why it has been so important to feminism to establish the mindful, social nature of mothering.

But just as women are refusing to be sacrificed, nonhuman nature requires more attention; manifested as the ecological crisis, it is revolting against human domination. Part of the resistance to contemporary feminism is that it embodies the return of the repressed, those things men put away to create a dualistic culture founded on the domination of nature. Now, nature moves to the center of the social and political choices facing humanity.

It is as if women were entrusted with and kept the dirty little secret that humanity emerges from nonhuman nature into society in the life of the species and the person. The process of nurturing an unsocialized, undifferentiated human infant into an adult person—the socialization of the organic—is the bridge between nature and culture.

The Western, male, bourgeois subject then extracts himself from the realm of the organic to become a public citizen, as if born from the head of Zeus. He puts away childish things. Then he disempowers and sentimentalizes his mother, sacrificing her to nature. The coming of age of the male subject repeats the drama of the emergence of the *polis*, which is made possible by the banishing of the mother—and with her the organic world. But the key to the historic agency of women with respect to nature vs. culture dualism lies in understanding that the mediating traditional conversion activities of women—mothering, cooking, healing, farming, and foraging—are as social as they are natural.

The task of an ecological feminism is discovering a genuinely antidualistic, or dialectical, theory and practice. No previous feminism can address this problem adequately from within the framework of its theory and politics, hence the necessity for ecofeminism. Rather than succumb to nihilism, pessimism and an end to reason and history, we seek to enter into history, to create a genuinely ethical thinking—where one uses mind and history to reason from what "is" to what "ought" to be and to reconcile humanity with nature, within and without. This is the starting point for ecofeminism.

Each major contemporary feminist theory—liberal, social, and cultural—has examined the relationship between women and nature. Each in its own way has capitulated to dualistic thinking, theoretically conflating a reconciliation with nature by surrendering to some form of natural determinism. As I have demonstrated, we have seen the same positions appear again and again in extending the natural into the social (cultural feminism), or in severing the social from the natural (socialist feminism). Each of these directions are two sides of the same dualism, and from an ecofeminist perspective both are wrong in that they have made a choice between culture and nature. This is a false choice, leading to bad politics and bad theory on each side, and we need a new, dialectical way of thinking about our relationship to nature to realize the full meaning and potential of feminism, a social-ecological feminism.

Absolute social constructionism on which socialist feminism relies is disembodied. The logical conclusion is a rationalized, denatured, totally deconstructed person. But socialist feminism is the antisacrificial current of feminism, with its insistence that women are social beings, whose traditional work is as social as it is natural. The fidelity to the social aspects of women's lives found in socialist feminism makes a crucial contribution to feminism.

It is for ecofeminism to interpret the historical significance of women's position at the biological dividing line where the organic emerges into the social. It is for ecofeminism to interpret this fact historically and to make the most of this mediated subjectivity to heal a divided world. The domination of nature originates in society and therefore must be resolved in society. Therefore it is woman as the embodied social historical agent, rather than a product of natural law, who is the subject of ecofeminism.

But the weakness of socialist feminism's theory of the person is serious from an ecofeminist standpoint. An ecological feminism calls for a dynamic, developmental theory of the person—male and female—who emerges out of nonhuman nature, where difference is neither reified or ignored and the dialectical relationship between human and nonhuman nature is understood.

Cultural feminism's greatest weakness is its tendency to make the personal into the political, with its emphasis on personal transformation and empowerment. This is most obvious in its attempt to overcome the apparent opposition between spirituality and politics. For cultural feminists, spirituality is the heart in a heartless world, whereas for socialist feminists it is the opiate of the people. Cultural feminists have formed the "beloved community" of feminism—with all the power, potential, and problems of a religion. For several years spiritual feminism has been the fastest-growing part of the women's movement, with spirituality circles often replacing consciousness-raising groups as the place that women meet for personal empowerment.

As an appropriate response to the need for mystery and attention to personal alienation in an overly rationalized world, it is a vital and important movement. But by itself it does not provide the basis for a genuinely dialectical ecofeminist theory and practice, which addresses history as well as mystery. For this reason, cultural-spiritual feminism (sometimes even called "nature feminism") is not synonymous with ecofeminism in that creating a gynocentric culture and politics is a necessary, but not sufficient, condition for ecofeminism.

Healing the split between the political and the spiritual cannot be done at the expense of the repudiation of the rational, or the development of a historically informed, dynamic political program. Socialist feminists have often mistakenly ridiculed spiritual feminists for having "false consciousness" or being "idealist." Socialism's impoverished idea of personhood, which denies the qualitative dimensions of subjectivity, is a major reason socialism—including socialist feminism—has no political base.[7] But many practitioners of feminist spirituality have eschewed thinking about politics and power, arguing that personal empowerment is in and of itself a sufficient agent of social transformation.

Both feminism and ecology embody the revolt of nature against human domination. They demand that we rethink the relationship between humanity and the rest of nature, including our natural, embodied selves. In ecofeminism, nature is the central category of analysis. An analysis of the interrelated dominations of nature—psyche and sexuality, human oppression, and nonhuman nature—and the historic position of women in relation to those forms of domination is the starting point of ecofeminist theory. We share with cultural feminism the necessity of a politics with heart and a beloved community, recognizing our connection with each other and with nonhuman nature. Socialist feminism has given us a powerful critical perspective with which to understand, and transform, history. Separately, they perpetuate the dualism of mind vs. nature. Together they make possible a new ecological relationship between nature and culture, in which mind and nature, heart and reason, join forces to transform the systems of domination, internal and external, that threaten the existence of life on earth.

Practice does not wait for theory—it comes out of the imperatives of history. Women are the revolutionary bearers of this antidualistic potential in the world today. In addition to the enormous impact of feminism on Western civilization, women have been at the forefront of every historical and political movement to reclaim the earth. A principle of reconciliation, with an organic practice of nonoppositional opposition, provides the basis for an ecofeminist politics. The laboratory of nonoppositional opposition is the actions taken by women around the world, women who do not necessarily call themselves feminists.

For example, for many years in India poor women who came out of the Gandhian movement have waged a nonviolent campaign for land reform and to save the forest, called the Chipko (the Tree Hugging Movement), wrapping their bodies around trees as bulldozers arrive. Each of the women has a tree of her own she is to protect—to steward.[8] When loggers were sent in, one of the Chipko leaders said, "Let them know they will not fell a single tree without the felling of us first. When the men raise their axes, we will embrace the trees to protect them."[9] These women have waged a remarkably successful nonviolent struggle, and their tactics have spread to other parts of India. Men have joined this campaign, though it was originated and continues to be led by women. Yet this is not a sentimental movement; lives depend on the survival of the forest. For most of the women of the world, interest in the preservation of the land, water, air, and energy is no abstraction but a clear part of the effort to simply survive.

The increasing militarization of the world has intensified this struggle. Women and children make up 80 percent of war refugees. Land they are left with is often burned and scarred in such a way as to prevent cultivation for many years after battle, so that starvation and hardship follow long after the fighting has stopped.[10] And here too, women—often mothers and farmers—respond out of necessity. They become the guardians of the earth in an effort to eke out a small living on the land to feed themselves and their families.

Other forms of feminist activism also illuminate an enlightened ecofeminist perspective.[11] Potentially, one of the best examples of an appropriately mediated, dialectical relationship to nature is the feminist health movement. The medicalization of childbirth in the first part of this century, and currently, the redesign and appropriation of reproduction both create new profit-making technologies for capitalism and make heretofore natural processes mediated by women into arenas controlled by men. Here women offered themselves up to the ministrations of experts,[12] internalizing the notion they do not know enough and thus surrendering their power. They also accepted the idea that the maximum intervention in and the domination of nature is an inherent good.

But since the onset of feminism in the 1960s, women in the United States have gone quite a way in reappropriating and demedicalizing childbirth. As a result of this movement, many more women want to know all their options and to choose invasive medical technologies only under unusual and informed circumstances. They do not necessarily reject these technologies as useful in some cases, but they have criticized motivations of profit and control in their widespread application. Likewise, my argument is not that feminism should repudiate all aspects of Western science and medicine. It is to assert that we should develop the knowledge to decide for ourselves when intervention serves our best interest.

But even the women's health movement has not realized a full ecofeminist perspective.[13] It has yet to grasp fully health as an ecological and social rather than individual problem, in which the systematic poisoning of environments where women live and work is addressed as a primary political issue. Here the community-based movements against toxic waste, largely led by women, and the feminist health movement may meet.

A related critical area for a genuinely dialectical practice is a reconstruction of science, taking into account the critique of science advanced by radical ecology and feminism.[14] Feminist historians and philosophers of science are demonstrating that the will to know and the will to power need not be the same thing. They argue that there are ways of knowing the world that are not based on objectification and domination.[15] Here again, apparently antithetical epistemologies, science and mysticism, coexist. We shall need all our ways of knowing to create life of this planet that is ecological, sustainable, and free.

As feminists, we shall need to develop an ideal of freedom that is neither antisocialist nor antinatural.[16] Ecofeminism is not an argument for a return to prehistory. The knowledge that women were not always dominated and that society was not always hierarchical is a powerful inspiration for contemporary women, so long as such a society is not represented as a natural order apart from history, to which we will inevitably return by a great reversal.

From an ecofeminist perspective, we are part of nature, but neither inherently good or bad, free or unfree. There is no one natural order that represents freedom. We are *potentially* free in nature, but as human beings that freedom has to be intentionally created by using our understanding of the natural world of which we are a part in a noninstrumental way. For this reason we must develop a different understanding of the relationship between human and nonhuman nature, based on the stewardship of evolution. To do this we need a theory of history where the natural evolution of the planet and the social history of the species are not separated. We emerged from nonhuman nature, as the organic emerged from the inorganic.

We thoughtful human beings must use the fullness of our sensibility and intelligence to push ourselves intentionally to another stage of evolution, one where we will fuse a new way of being human with a sense of the sacred, informed by all ways of knowing—intuitive *and* scientific, mystical and rational. It is the moment when women recognize ourselves as agents of history—even unique agents—and knowingly bridge the classic dualism between spirit and matter, art and politics, reason and intuition. This is the potentiality of a *rational* reenchantment. This is the project of ecofeminism.

The domination of nature is inextricably bound up with the domination of persons, and both must be addressed, without arguments over the primary contradiction in search of a single appropriate point for revolution. There is no such thing. And there is no point in liberating people if the planet cannot sustain their liberated lives, or in saving the planet by disregarding the preciousness of human existence, not only to ourselves but to the rest of life on earth.

NOTES

1. One of the major issues at the United Nations Decade on Women Forum held in Nairobi, Kenya, in 1985 was the effect of the international monetary system on women, and the particular burdens women bear because of the money owed the First World, particularly U.S. economic interests, by developing countries.
2. See Peter Singer, *Animal Liberation: A New Ethics For Our Treatment of Animals* (New York: Avon Books, 1975).
3. It is one of the absurd examples of newspeak that the designation "pro-life" has been appropriated by the militarist right to support forced childbearing.
4. For a fuller discussion of this point, see William Leiss, *The Limits to Satisfaction: An Essay on the Problem of Needs and Commodities* (Toronto: Univ. of Toronto Press, 1976).
5. In *German Ideology,* Marx cut his teeth on the "natural order" socialism of Feuerbach, although he had tended toward a "naturalistic socialism" himself in his early *Economic and Philosophic Manuscripts.* See T. B. Bottomore, *Karl Marx: Early Writings* (New York: McGraw-Hill, 1964).
6. On the social, mindful nature of mothering see the work of Sara Ruddick, especially "Maternal Thinking," *Feminist Studies* 6 (Summer 1980): 342-67; and "Preservative Love and Military Destruction: Some Reflections on Mothering and Peace," in Joyce Trebilcot, ed., *Mothering: Essays in Feminist Theory* (Totowa, N.J.: Rowman & Allenheld, 1983), 231-62.
7. The most vital socialism in the world today is liberation theology, with its roots in the Catholic- based communities of the poor in Latin America.
8. Catherine Caufield, *In the Rainforest* (Univ. of Chicago Press, 1984),154-156.
9. Ibid.,157.
10. See Edward Hyams, *Soil and Civilization* (New York: Harper & Row, 1976).
11. West German Green Petra Kelly outlines a practical, feminist, green political analysis and program, with examples of ongoing movements and activities, in her work. See Petra Kelly, *Fighting for Hope* (Boston: South End Press, 1984).
12. See Barbara Ehrenreich and Deidre English, *For Her Own Good* (New York: Doubleday/Anchor, 1978).
13. I am indebted to ecofeminist, sociologist, and environmental health activist Lin Nelson for pointing out to me why the feminist health movement has yet to become ecological.

14. See Elizabeth Fee, "Is Feminism a Threat to Scientific Objectivity?" *International Journal of Women's Studies* 4(1981). See also Sandra Harding, *The Science Question in Feminism* (Ithaca, N.Y.: Cornell Univ. Press, 1986), and Evelyn Fox Keller, *Reflections on Gender and Science* (New Haven: Yale Univ. Press, 1985).
15. See Evelyn Fox Keller, *A Feeling for the Organism: The Life and Work of Barbara McClintock* (San Francisco: W. H. Freeman, 1983).
16. The crosscultural interpretations of personal freedom of anthropologist Dorothy Lee are evocative of the possibility of such an ideal of freedom. See Dorothy Lee, *Freedom and Culture* (New York: Prentice-Hall, 1959).

and discourses become irrelevant, divorced from real-life experience. For example, many people now recognize the implications resulting from the conveniences of technological advances in chemicals used in agriculture and manufacturing and the hazards they pose to public health and the environment. Thus may the consumer culture have its limits. However, the connections between racial discrimination, class inequities, and ecological inequities may not be so readily perceived.

Under these conditions, how can the voices and the organizing capacities of communities struggling for environmental justice be strengthened and mobilized? How can communities generate popular support for common goals and a society not based on market values in a world dominated by large corporations?

In a more culturally pluralistic world, the opportunities for collaboration and innovative common practices abound. Realizing these possibilities will require the development of cultural strategies and skills necessary to negate corporate myths and symbolic activity that surround debates about the environmental crisis. All citizens can develop these skills. Everyone has a story to tell.

How will people imagine a society that transcends racism, sexism, and class? If environmental justice means equitable, cooperative relations among ourselves and everything else on the planet, then social transformation will mean rethinking these relations. Creating such visions is necessary for survival. The conditions of struggle require a politics that challenges principles of growth, consumerism, and corporate power that will not permit the burden of ecological devastation to be borne by people of color and disenfranchised populations.

THE CRISIS OF MEANING

Public debates about the environment often ignore central connections between ecology and race, class, and gender. Political success may depend on exposing the way environmental issues get limited to technical discussions about acceptable risk and scientific uncertainty. For example, how do corporations create acceptance of the idea that we are all responsible for the environmental crisis, when the source of most environmental degradation derives from corporate and military decision making? Environmental debates often use the language of risk assessment, cost-benefit analysis and trade-offs that ignore social costs. Everyday environmental imagery from the corporations promotes more waste and consumption through symbols of progress, adventure, the future, and affluence. Consider, for example, automobile commercials, which ignore almost all social context and public consequences (e.g., traffic, large junkyards for inoperable cars, and ecological consequences).

Large corporations appropriate green symbols in marketing products and advocate changing individual consumption patterns, instead of production and investment processes. For example, companies have spent millions of dollars opportunistically promoting environmental awareness among customers, making unprovable claims with terms like "degradable" and otherwise treating the ecological crisis as a public-relations crisis. They describe environmental hazards as the result of technological failures, mistakes, or accidents that suggest uncontrollable or random forces, as exemplified by the Exxon *Valdez* oil spill being blamed on the ship's captain and otherwise presented as an accident. These corporations endlessly offer tips to consumers that focus on what each individual can do: become a "green" consumer of their products.

Another approach may be to think of replacing a society founded on accumulation of social wealth for private purposes with one where citizens determine the use of the social wealth in common. The core features of a reconstructed view include a challenge to absolute property rights, a belief that a clean environment is a fundamental right, collective bargaining for citizens, and a recognition that democracy is about decisions that affect our communities—in production, investment, and the use of resources. But how does a particular perspective become embedded in daily life? How can people reaffirm their history, culture, and noncommercial values when corporate culture shapes so much of everyday life?

The power of corporations' symbolic activity partly results from the concentration of economic power, the control of production and resources, the means of communication, and the management of information and images—not because of any conspiracy but by how these institutions that govern our lives unconsciously influence the language and categories in which people think about the world. For example, the Dow-Jones Industrial Average is a dominant social statistic in the United States. Newspapers offer no labor section, only a business section. These taken-for-granted forms and concepts constitute a powerful ideological force. Such power translates into social and cultural power to determine what people will see, read, hear, and experience, obscuring the sources of environmental crises and the underlying political interests. Defining the environment as external from other social forces and relationships, such as poverty, racism, and colonialism, makes social reconstruction difficult.

People sometimes cannot resist an intrusive, powerful information system that constantly bombards them with implicit messages and stories unrelated to their lives. Increasingly, as people have become more dependent on products, ideas, and technologies outside their own traditions and local experience, they can become susceptible to corporate interpretations of reality. For example, a supermarket is more than just a place to buy groceries; it is a theater of images and displays. Supermarkets design their space to create an ambiance and mood that will stimulate sales on impulse, beyond what the customer intended to purchase. Domination works most effectively when it is not seen as domination—when the categories in which we think appear as natural or the result of random events. The corporate perspective is powerful partly because it is organized, with many vehicles of expression. However, the destruction of traditional bases for social meaning and the detachment of corporate culture from real needs always provides openings for reclaiming critical faculties. The way people invest meaning in their work and identities often presents oppositional possibilities. Yet they sometimes require more conscious activity to create an effective political culture.

How do grass-roots organizers struggling against hazardous waste in their communities express a vision of a less-toxic society? How will they challenge the representation of the ecological as external to issues of social justice? Because culture is always in a process of development, never forcibly imposed and constantly shifting terrain, opportunities exist for disrupting the deluge of corporate representations.

DEVELOPING A MOVEMENT

Many of those working to build a multiracial, multi-issue movement for environmental justice are effectively incorporating democratically developed forms of cultural expression as an essential component of community development. The expressive power of people in their everyday life experiences is a way to communicate a reconstructed vision of society opposing the cultural institutions and symbols of corporate capital. The practices are already in use, even if disconnected from their own authority and power in the face of the commercial media. As activists draw out alternative visions, they can unleash everyone's imaginations, so that people can notice the value of what they sometimes do not reflect upon in their own self-expression as communities. The objective of cultural activism is not simply education but changing consciousness—questioning corporate cultural stories and symbols in a way that leads to discovery and creative thinking—just as the women's movement exposed how taken-for-granted language, imagery, and behavior functioned as tools to oppress women.

In challenging basic assumptions about the organization of society, using better facts and logic, exposing scandals, or presenting a better message cannot itself support an effective social movement. Getting media attention is insufficient because the media absorbs, coopts, and reframes perspectives outside its narrow boundaries of acceptable discourse. Many advocates reject the standard narratives and rhetoric that frame the meaning of ecological dangers posed by the nuclear industry. Since everyone processes and registers more than information on their consciousnesses, education, environmental literacy, and scientific knowledge may be inadequate without political literacy and a method for defining, deconstructing, analyzing, and reconstructing alternate perspectives.

Strengthening the movement for environmental justice, activists are incorporating environmental issues into a broader agenda for social justice. This means making common cause with labor, civil rights, women, peace, and public-health movements on an international basis, overcoming language and cultural differences. The environmental crisis is potentially a powerful unifying phenomenon that can link seemingly separate issues and peoples.

Cultural activism is a form of political action practiced by many grass-roots groups that serves as a force for change in that struggle. It represents a way of giving voice to people in their own language and images, derived from historical memory and current experience.

CULTURAL ACTIVISM

The term "culture," as used here, broadly refers to the history, tradition, rituals, customs, lived conditions, and expressive life of people or communities. Always in flux and filled with contradictions, it includes the writings, music, paintings, songs, stories, oral histories, street art, games, and dramas of a people that are significant to them. More than the sum of these manifestations, culture articulates human subjectivity, meaning, and a people's presence and identity in history. It represents the way a community of people reflects on and represents itself.

Politics always serves as a theme, a narrative in people's lives. And almost all political activity is a cultural expression. Political demonstrations, for example, are theaters in which everyone is an actor. Culture is not the sole province of arts agencies, museums, galleries, or artists. It is a living, breathing, historical, social process in which everyone plays a role. Historically, people have preserved their collective knowledge through storytelling, music, and cultural documentation. Thus, everyday cultural life is an important locus for social and political conflict and change.

Cultural activism is a strategy for social change and liberation, challenging basic assumptions about society, transforming political power, and building political unity. It offers a way for people to reflect on their relationship to daily life and to own and control their images and representations.

Cultural activism can ignite critical consciousness and interrogate or disrupt society's systems of power. As a method for revitalizing social imagination, cultural activism can produce new meanings and social theories for envisioning a reconstructed society, affirming unofficial versions of history and initiating a different path toward human and social development. Cultural activism is not an appendage or entertainment, but is at the core of movement organizing and community. The civil rights movement, the struggles for women's rights, and the fight against AIDS are strong examples of the uses of cultural expression, particularly music and street theater. It can also be a core feature of organizing for environmental justice, by developing artistic, organizing, cooperative, and leadership skills for social change. Its practices include expressing community identity, history, visions of the future, and aspirations of different peoples through their own symbols, language, and stories, in a socially conscious, public way rooted in social life. Such expression remains distinct from the narratives in the media and textbooks that, through endless repetition, come to appear as natural rather than socially constructed, such as Columbus's "discovery" of America.

Cultural activism builds solidarity and political awareness among communities with common concerns, sometimes through oppositional activity. Jane Sapp, a musician and cultural activist, describes cultural activism as a way of developing community by connecting diverse people and making them feel they can act to build a movement, rather than function as spectators. In her words, "facilitating a community's appreciation and use of its culture can be a powerful tool of building community and empowerment.... Culture affirms and allows people to speak and act from their strongest and most firmly rooted context."[2]

Cultural expression, an important component of community development, can enable citizens to take control of the direction of their communities and build self-esteem. Taller Puertorriqueño in Philadelphia, for example, a "community-based art cultural-educational organization dedicated to the preservation and development of Puerto Rican and Latin American artistic and cultural traditions," believes that "everyone in the community has a right to create art," which means renovating the neighborhood as well as developing artistic creativity.[3] Taller's political significance lies in giving the community a voice for documenting itself and its needs, and helping people connect with each other.

The CAMEO (Community, Autobiography, Memory, Ethnography, Organization) Project, a collaboration between the New Museum in New York City, the City University of New York, and local immigrant community groups, explores recent Latino American immigration to New York City in three neighborhoods as a means of building community. With artifacts, storytelling, oral histories, art installations, and other techniques that examine neighborhood life in and outside the Museum, the project addresses cultural transition, resistance to assimilation, and the role of art in community organizing. The project is "being developed as a model for collaborative community outreach,"[4] whereby academics work with community residents to collect histories and present them to the public. Some of the works explore problems in building community leadership, while others consider the transformation of traditional cultures. The project has been a satisfying, rewarding experience for many of the participants.

As an ongoing experiment, this project considers the ways in which the Museum can function as a forum for different constituencies in the community to further their goals as they define them. Museums as repositories of elite culture rarely serve this function. A central issue for the project is representation: Can the culture of everyday life be brought into the Museum without distortion or exploitation? How can the Museum engage in political issues and transcend an elitist orientation?

Many projects are created where struggles take place—not in museums and galleries but in hospitals, prisons, schools, and the streets. Some work toward developing cultures of resistance and a different type of social imagination. Activists working in this way act as social change agents and analysts; they need to be understood as equal partners in organizing for social justice. They work to generate creativity and resistance through a collective process. Some donate their work to communities; others seek to achieve participatory culture by animating everyone to collective expression. Whatever the approach, making the links between labor activism, feminist activism, or environmental racism demands collaborative multicultural projects.

What approaches do cultural activists use? Many dramatize and expose injustice to reinforce the voices of those subject to injustice, set new agendas, and sharpen understanding of alternative actions and their consequences. Their role is to clarify and expose power relations, transforming society by creating a common vision among diverse cultures. More important, many cultural activists involve the community in their work and place artistic imagination at the service of communities.

The murals of the Chicago Public Art Group, for example, are sponsored by labor organizations, neighborhood groups, and social-service agencies. One of their goals, beyond the content of their work, is to suggest a public or communal ownership of the space in which their work appears. More than this, they attempt to minimize differences between artists and ordinary people who brainstorm and work with artists on design of the murals. By developing close, equitable relations between artists and community residents, they are able to create multiple visions in the process. Because their work is collectively produced and rooted in community conditions, it serves as a public articulation of community experience and a way people give meaning to their lives.

Cultural activists strive to make their work accessible and relevant. They believe in informing the public about the issues on which they center their work. Community arts activists provide training to their constituents in reading the dominant culture and using artistic practices to document their common concerns, histories, and identities.

How does cultural activism support social change? How can it be strategically and aesthetically effective at the same time? If people take control of their image, history, and language, they can more readily take control of their lives. For example, the Artist and Homeless Collaborative, a public-art project founded by Hope Sandrow in New York City, offers homeless women residing in city shelters the chance to study and produce art, individually and as a community. By developing self-expression and fostering collaboration, these women have created artworks for exhibition and engaged in public discussions about their work. The skills these women have learned have fostered personal growth, curiosity, and confidence, and have helped enable them to survive.

The project shows the relevance of self-expression to political activism, particularly the way in which the homeless create their own images of homelessness. With these kinds of skills, they are no longer dependent on experts and mass media to represent them. Speaking through theater, murals, performance, or video is a powerful means for gaining control of their own symbols and language.

Pregones Theater in the Bronx, New York, produced the Embrace Residency Project as a means to educate the Latino American community about AIDS and to prevent its spread. This project emphasizes the creation and performance of innovative and challenging theater rooted in Puerto Rican traditions and popular artistic expressions, with the goal of offering the community an artistic means to question, reaffirm, and enhance its role in society. What is distinctive about this project from the point of view of cultural activism is its collaboration with Bronx community-based organizations concerned with AIDS education and its practice of involving audiences in formulating plans to engage in activities that will reduce their risk of infection. Working closely with service providers, teachers, and health aides, Pregones staff conduct workshops using theater techniques to improve these community workers' skills when talking about AIDS.

An important objective of cultural activism is to expose what cultural analyst Frederic Jameson calls the "political unconscious."[5] Cultural activism is a way of surfacing that which has been historically repressed. Similar to the way that a therapist exposes the individual unconscious by enabling people to reexperience what they already know and have always known, the cultural activist works to create a collective consciousness capable of producing a public agenda. This means recovering lost history and making it accessible.

By incorporating cultural activism into their work and collaborating with art makers, social activists find that they can improve their organizing abilities. Increasingly, many art makers are striving to link their work more directly with social movements. They usually view effective and progressive art work or cultural expression as more than a spectacle; it must be provocative and confrontational. It must provide an idea of resistance while giving voice to communities that often lack a voice.

Community arts activists working for social change may find the resources and support they need through building direct links with social movements and political organizations. For example, Sally Jacques, a performance artist in Austin, Texas, created the 64 Beds Project; to dramatize homelessness, sixty-four artists created beds for an all-night street performance. Participants slept on them and recorded their thoughts about bed and home to encourage people to consider their relationship to the homeless. Homeless people or representatives from their organizations usually speak or perform at the event. Jacques, who performed the Project many times in cities across the country, generally works with familiar objects or images with ritual, music, and dance to remake what we see in a sharpened, different perspective. Most important, the public is physically involved in the performance segment, making it a participatory experience.

The spirit of cultural activism is producing a new generation of creative organizers who are applying their skills to environmental justice.

CULTURAL ACTIVISM AND ENVIRONMENTAL JUSTICE

The creation and reaffirmation of community culture can advance grass-roots organizing for environmental justice. Identifying shared history is a way of building community solidarity. Organizers search for autonomous spaces for action outside regular channels such as the mass media, elections, or other bureaucratic institutions that narrow the boundaries of legitimate political action.

The last decade has given rise to numerous models for action on environmental justice. Some seek to mobilize support on specific issues. Some present information in appealing and graphic ways. Others seek to create community. What may finally be most effective in shaping consciousness are those features of everyday life and popular culture that incorporate environmental justice themes and images. Such expressions are usually opposed to and counter the themes presented by multinational corporations about commodities, value, markets, and growth in everyday consciousness. What many of the activists and organizations described here have in common is the public, accessible character of their work, rooted in people's experiences. Their methods are multimedia, replicable in other places, and usually inexpensive.

Appalshop Media Arts Center in Whitesburg, Kentucky, produces work in film, video, audio, theater, and photography, celebrating the culture and concerns of people from Appalachia. As a way to preserve indigenous culture, they train people in the use of different media. One project, a film called "Chemical Valley," produced for the P.O.V. (Point Of View) series on public television explores the experiences of Union Carbide workers in Institute, West Virginia who were involved in a serious chemical accident at their plant. Mimi Pickering, who produced and directed the film, made extensive contacts with environmental justice organizations to publicize the film, and made it available to schools and universities. Appalshop produced a resource guide about the issues. They also involved community leaders in evaluating the film and its supporting materials. Appalshop's general goal is to develop a network of environmental justice activists and community organizations that use media as an organizing tool. "Chemical Valley" was thus a cooperative project produced in political struggle and made available to those who needed it.

Independent video producer Branda Miller created a ninety-minute video about the lives of individuals victimized by nuclear radiation, and toxics from chemical plants, who became transformed into activists. The film depicted three locations: the Hanford nuclear plant in Washington State, Latino American communities in the San Joaquin Valley, and "Cancer Alley" in rural Louisiana. As an art maker, Miller's goal is to empower people by giving them a chance to express themselves by controlling the essential features of the project. She transfers her knowledge of media directly to the subjects of the video and provides them the opportunity to determine its content and structure, unlike most documentaries where subjects are only actors. In this video, Miller sought out community activists representing diverse communities and asked them to participate actively in the work. She encouraged them to speak the truth in public. From the early planning stage to editorial work and distribution, the subjects controlled their own images. They, after all, were the experts and spokespeople for their communities. Using an innovative cultural-activist approach, Miller offered the subjects of the video control over their sequence, a process that built trust and engaged the participants. Thus, they could determine if they wanted to direct the sequence, write, edit a scene, or experiment with an idea. Most important, they set the standards. Miller's role was to realize their vision as effectively as possible. While such collaborative work can prove difficult, it resonates with authenticity in a way that more conventional documentaries do not. For Miller, video is a change-making tool.

At the rough-cut viewing, about a dozen people connected with the project—including Native Americans, African Americans, whites, and whistle-blowers—used the screening as a forum for debate on the environment. Many of these people, who had never worked together or even known about each other, began to form alliances. Instead of showing the video primarily to viewers of public television, it will be promoted and distributed so that other communities with similar problems will have access to it.

Toxic Avengers, an organization of Latino American and African American high-school and college students in Brooklyn, New York, uses indirect action, marches, teach-ins, workshops, and civil disobedience at toxic waste sites in a heavily polluted community. They employ the tactics of the civil-rights movement, including sit-ins, rallies, and demonstrations. Toxic Avengers recognize the importance of cultural animation and train community residents in their neighborhood and elsewhere around the country in effective cultural strategies. With wall posters and demonstrations, they waged a significant campaign with hundreds of participants against the Radiac Research Corporation, which transports and stores toxic wastes.

San Francisco playwright and poet Cherie Moraga's widely acclaimed play, "Heroes and Saints," is a story of resistance about one farm worker family and its devastating experience with pesticides. Moraga relies on images that connect art to reality and portray people as activists rather than as victims. A distinctive feature of her work is the way in which the entire production incorporates elements that support organizing. Presentations of her play include speakers and written materials and are directed to audiences who can especially benefit from the work. For example, a group fighting toxics in Kettleman City, California, bought the entire house for two shows so that they could effectively mobilize their community and support their petition drive. In her view, using theater as an organizing tool must be balanced by allowing the aesthetic to work at its own level: "Political theater must be visionary in order to anticipate a social movement, as opposed to presenting what people already understand."

The Farmworkers Support Committee in Glassboro, New Jersey, uses participatory theater to work with farm workers on the dangers of pesticides. Based on the popular education techniques of Paulo Friere, these productions enable farm workers to examine themes central to their community based on their own expertise and knowledge of the issues. One of the productions uses a "Wheel of Fortune" game-show format to enable the group to engage in dialogue about their work, using familiar cultural symbols, and to develop collective solutions.

CONCLUSION

Organizing for environmental justice and against the social inequities that produce injustice is a process that opposes fragmentation and divisiveness central to U.S. culture. In the coming struggles to sustain communities, everyone can be a creative participant; everyone must have a part.

Building a vital, progressive community requires awakening people to their cultural roots and recognizing the opportunities for creative political expression. Stronger collegial bonds between art makers and organizers for environmental justice, within the movement, can strengthen it and expand its meaning for a wider public. Environmental justice organizers can move their organizing toward a more self-conscious form of direct action through cultural activism. The elimination of the distinctions that separate art and cultural expression from the rest of life is crucial to success. By enabling people to express their own lives and imagine a different society, citizens can make the mass media irrelevant and create their own visions of the future.

NOTES

1. See the special issue of *Advertising Age*, "The Green Marketing Revolution," 29 January 1991.
2. Descriptive document from the Center for Cultural and Community Development, n.d.
3. Taller Puertorriqueño/The Puerto Rican Workshop, Document of History and Purpose., n.d.
4. CAMEO Project Description, New York, 1992.
5. See Frederic Jameson, *The Political Unconscious: Narrative as a Socially Symbolic Act* (Ithaca, N.Y.: Cornell Univ. Press, 1981).

Part Two

Practice and Politics:

Ecological Inequities and Visions of Possibility

9

A Society Based on Conquest Cannot Be Sustained

Native Peoples and the Environmental Crisis

Winona LaDuke

BOTH WORLDWIDE AND in North America, Native people are at the center of the present environmental and economic crisis. This is no coincidence. Five hundred years after Columbus's invasion, Native people still maintain a significant presence on the North American continent, and in the Western hemisphere overall. Although an estimated two thousand Native communities have become extinct in the past four centuries, over seven hundred Native communities remain on the continent, two hundred in Alaska alone, eighty in California, and hundreds more scattered as islands in a sea of what is called the United States and Canada, retaining lands reserved by treaty or other agreement.

While demographically Native people represent a minority population in North America, we maintain land occupancy over substantial areas of the continent. In many regions in the United States, we are the majority population—parts of New Mexico, Arizona, Northern Minnesota, the Dakotas, and Montana are examples. But even more striking are the statistics for the Arctic and sub-Arctic.

From the fiftieth parallel north, in what is called Canada, the majority population is Native—85 percent or more. The Northwest Territories, for instance, is under legislative mandate to divide the land into two Native territories, one Dene and one Inuit. Not surprising, since almost everyone there is Native. But elsewhere across the north—northern Quebec, Newfoundland, Labrador, and Ontario—all the way to British Columbia, the northern population is decidedly Native. The Native population is the majority population in the upper two-thirds of Canada, or about one-third of the continent. Within this context Native thinking, the survival of Native communities, and the issues of sovereignty and control over natural resources become

central to North American resource politics and the challenge for North A
of conscience. Consider these facts:

- Over fifty million indigenous peoples inhabit the world's remaining rain forests;
- Over one million indigenous people will be relocated to allow for the development of hydroelectric dam projects in the next decade;
- The United States has detonated all its nuclear weapons in the lands of indigenous people, over six hundred of those tests within property belonging to the Shoshone nation;
- Two-thirds of all uranium resources within the borders of the United States lie under Native reservations—in 1975, Indians produced 100 percent of all federally controlled uranium;
- One third of all low-sulphur coal in the western United States is on Indian land, with four of the ten largest coal strip mines in these same areas;
- Fifteen of the current eighteen recipients of nuclear-waste research grants, so-called monitored retrievable nuclear storage sites, are Indian communities; and
- The largest hydroelectric project on the continent, the James Bay project, is on Cree and Inuit lands in northernCanada.

In order to explore the consequences of U.S. colonial occupation for the health, sovereignty, culture, and environment of Native peoples, I will briefly discuss three subjects: the relationship of indigenous values to sustainable communities and economies, the legacy of colonialism, and the present political and economic circumstances of Native North America, with an emphasis on North American energy issues, Native resistance, and the rebuilding of Native communities.

SUSTAINABLE COMMUNITIES: MINO BIMAATISIIWIN

The key to a sustainable society is accountability to natural law. Indigenous, or land-based societies (wherever they are found in the world or in history) understand that all life is accountable to natural law. Laws made by nations, states, provinces, cities, etc. are inferior to this supreme law. In the Anishinabeg community, we have a value system and code of behavior/ethics that keeps communities and individuals in line with natural law.

Mino bimaatisiiwin (the "good life") is a central value to the Anishinabeg people who, to this day, occupy a great portion of the North American continent. *Mino bimaatisiiwin* guides the way of life and is the essence of sustainability on this land. An alternative interpretation of the word is "continuous rebirth." Two basic tenets essential to living within natural law and within *mino bimaatisiiwin* are cyclical thinking and reciprocal relations with the earth.

Cyclical thinking, common to most indigenous or land-based cultures and value systems is an understanding that the world, time, and all parts of the natural order flow in cycles—whether those cycles belong to the moon, the tides, our bodies, seasons, or life itself. Within this understanding is a clear sense of birth and rebirth, and a knowledge that what one does today will affect one in the future, on the return.

A second concept, reciprocal relations, defines the responsibilities and ways of relating between humans and the ecosystem. Simply stated, the resources of the ecosystem—whether wild rice or deer—are, with few exceptions, recognized as animate and, as such, gifts from the Creator. Thus, one could not take life without a reciprocal offering, usually tobacco *(saymah).* Within the practice of reciprocity is also an understanding that you take only what you need and leave the rest.

Implicit in the concept of *mino bimaatisiiwin* is a continuous inhabitation of place, an intimate understanding of the relationship between humans and the ecosystem, and the need to maintain the balance. These values and cultural tenets make it possible for many indigenous peoples to maintain economic, political, religious, and other institutions for generations in a manner which would today be characterized as sustainable.

By nature, a socioeconomic system based on these ascribed indigenous values must be decentralized, self-reliant, and very closely based on the land of that ecosystem. Not surprisingly, the nature of most indigenous economies constitutes a diversified mix of hunting, harvesting, and gardening, all balancing human intervention and care and all in keeping with the religious-cultural systems' reliance upon the wealth and generosity of nature. A nurturing relationship with the natural world is essential to indigenous societies.

The Anishinabeg or Ojibway nation represents an example of a sustainably based community. Within a region encompassing the southern parts of four Canadian provinces and the northern parts of five American states, this nation retains a common culture, language, history, governance, and land base, the five indicators, according to international law, of the existence of a nation of people. This nation functions within a decentralized economic and political system, with much of the governance left to local bands (similar to villages or counties), through clan and extended family systems. The vast natural wealth of this region and the resource management systems of the Anishinabeg, emphasizing sustained yield, have enabled people to prosper for many generations. This system provides sustenance for both domestic production and production of exchange for export. Whether the resource is wild rice or white fish, the extended family as a production unit harvests within a social and resource management code that ensures sustained yield.

Anishinabeg cultural practices, maintained for thousands of years, are indicative of the way of life in many native or land-based communities. This cursory overview represents a small window for viewing the context of indigenous values, economic systems, and their relevance in the present discussion of sustainable development.

CONQUEST: A WAY OF LIFE

In sharp contrast is the system of capitalism and other forms of industrialism that lack respect for people and their environments in an insatiable quest for resources. This is particularly obvious in the United States, which consumes one-third of the world's resources and hosts only 5 percent of the world's population.[1]

Columbus provided the entrée for this system into the Western hemisphere. The holocaust that subsequently occurred in the Americas is unparalleled on a world scale, and in its wake was the disruption necessary to destroy many indigenous economic and governmental systems. It is the most comprehensive system of imperialism ever witnessed by humanity. While no one knows exactly how many people were killed since Columbus's invasion, one conservative estimate suggests that the population of indigenous people in 1492 was 112,554,000 in the Western hemisphere and 28,264,000 in 1980.[2]

This genocide facilitated a subsequent process of colonialism that established new relations of dependency and underdevelopment between indigenous nations and colonial or settler nations in the Americas. Three elements define these relations. First, colonialism has been extended through an expansion of centralized power by the spread of Christianity, western science, and other forms of Western thought; the socioeconomic practice of capitalism; and the military-political practice of colonialism. The appropriation of land and resources from indigenous populations characterizing these relations results in their living in circumstances of material poverty and ill health.

As Ward Churchill and I wrote some time ago, "Land has always been the issue central to North American politics and economics. Those who control the land are those who control the resources... whether the resource at issue is oil, natural gas, uranium... water, agriculture or land ownership. Social control and all the other aggregate components of power are fundamentally interrelated."[3] This may be true everywhere, but the peculiarities in this hemisphere, with its imported apparatus of socioeconomic power, make the equation especially acute.

As a vast portion of the remaining natural resources in the North American continent still underlie Native lands, and as disposal of toxic and nuclear wastes on Indian reservations continue, the residual structures of colonialism are required to continue the extraction on unequal terms. In addition, Native communities are focal points for the excrement of industrial society, a situation made possible by both the colonial relationship of the United States and Canada to Native peoples and general conditions of environmental racism in each country. Any discussion of present environmental circumstances or the possibilities for sustainable development in a North American context must recognize these historical relations.

AN ENERGY POLICY THAT VIOLATES NATURAL LAW

I want to focus briefly on North American energy policy from the viewpoint of Native peoples by looking at two specific issues: the development of nuclear and hydroelectric energy.

North American energy policy is based on short-term capital gains and accelerated consumption, not long-term sustainability, much like society itself. Consequently, seemingly endless proposals of nightmarish proportions loom on the horizon. We have had more than our share, from lethal uranium mines, to coal gasification and slurry pipelines in virtual deserts, to the Frankenstein monster of all hydroelectrical projects, an untested ecological manipulation intended for American markets and an outmoded development plan for Quebec.

The nuclear age blew into Native America with the first uranium for atomic weapons. Next came thousands of uranium mines throughout the Navajo reservation (an area the size of West Virginia), today mostly abandoned but still emitting high levels of radioactive gases and endangering the community.

According to Victor Gillinsky of the U.S. Nuclear Regulatory Commission, "Uranium mining and milling are the most significant sources of radiation exposure to the public of the entire nuclear fuel cycle, far surpassing nuclear reactors and nuclear waste disposal...." A 1978 report by the Los Alamos Scientific Laboratory stated that, "perhaps the solution to the radon emission problem is to zone the land into uranium mining and milling districts so as to forbid human habitation."

The production of uranium, or yellowcake, from uranium ore, usually requires the discharge of significant amounts of water and the disposal of significant portions of radioactive material. Uranium mill tailings, the solid wastes from the uranium milling stage of the cycle, contain 85 percent of the original radioactivity in the uranium ore. One of these products, radium 226, remains radioactive for at least 16,000 years.

In 1975, 100 percent of all federally produced uranium came from Indian reservations. That same year there were 380 uranium leases on Indian lands, as compared to four on public and acquired lands. In 1979, there were 368 operating uranium mines in the United States. Worldwide, estimates are that 70 percent of uranium resources are contained on indigenous lands.[4]

Thirty years after it blew in, the nuclear age moved out, largely coinciding with an aggressive antinuclear movement and a subsequent drop in uranium prices. Only 120 reactors have been built in the United States compared to the one thousand reactors proposed for this country by the year 2000. Companies moved to cheaper uranium deposits on Saskatchewan Cree, Dene, and Metis lands, the lands of Australian aborigines, and many others. And America looked toward hydroelectric power for a cleaner electricity.

HYDROELECTRIC EXPLOITATION: JAMES BAY

James Bay, at the base of Hudson Bay, is the largest northern drainage system on the North American continent. Virtually every major river in the heartland empties there. This makes the bay a rich ecosystem teeming with wildlife, the staging ground for migratory birds, and a feeding area for the largest migratory herd of mammals on the continent, the George's River caribou herd. Approximately 35,000 Cree, Innu, Inuit, and Ojibway people live within the region and are dependent upon the ecosystem. The way of life is land-based: a subsistence, hunting, harvesting, and tourism economy from which at least 50 percent of the food and income for the region originates.

In 1972, the James Bay I project was introduced in northern Quebec as a joint activity among a number of public and private corporations. The intent was to produce 10,000 megawatts of electricity by putting 1150 square kilometers of land under water and behind dams. Ultimately the initial project concentrated along the East Main and Rupert Rivers and ruined the ecology of some 176,000 square kilometers, an area about the size of the former West Germany. The Native people did not hear of the project until planning was well underway. Most of the power was to be sold to companies in New York and New England.

After several years of litigation, the Quebec Court of Appeals ruled that too much money had already been spent on the project to abort it, even though no serious assessment of damage to the environment or Native people was ever authorized. Nor did the Court consider alternatives that would protect cultural and bioregional diversity. Basing their decision on the "balance of convenience" argument, the court ruled that the project should proceed. Within eleven months, 400 kilometers of paved road pushed into the heart of the territory. With the building of three large power stations and the flooding of five large reservoirs, four major rivers were destroyed: the LaGrande was drowned; the Eastmain and Opinaca were dried up, and the headwaters of the Caniapiscau were diverted to flow south into the LaGrande, rather than north into Ungava Bay. Five 735 KV (Kilovolt) power lines through Cree territory cut a swath through the wilderness all the way to the U.S. border. Six new power plants are planned to provide an additional 4500 megawatts of power. Because these and many more proposals are seen as upgrades, rather than as new projects, no environmental assessment is planned even now.

Under present conditions, the environmental impact is devastating. Mercury contamination of the LaGrande reservoir system is a by-product of the decay of vast amounts of vegetation in the ecosystem. Mercury levels at the reservoirs are six times above safe levels, and about two-thirds of the people downstream from the reservoirs have mercury contamination in their bodies—some at thirty times the acceptable level. The devastation of vast amounts of hunting and trapping territory resulted in economic and social dislocation from loss of food and cultural practices.

Even with widespread opposition to the dams, province officials propose a new phase for the economic development program. Over $62 billion is being raised on the bond market by Hydro Quebec to secure investments. If successful, the plan will flood an area roughly the size of Lake Erie and use the electricity to fuel energy intensive aluminum smelters and export contracts to the United States, primarily to New England.

Phase two of the James Bay Project is even more devastating. The first proposal calls for the construction of another $600 million worth of roads and airports. Numerous small rivers will be diverted into a large one, and eighty dams will be constructed. The cumulative impact of this phase will destroy caribou calving grounds and habitats for migratory birds, and will destabilize the water in an entire ecosystem. According to Jan Bayea of the Audubon Society, the entire ecosystem will be lost in fifty years from these two projects.[5] Even worse, their impact is compounded by fifty-eight smaller dams proposed for the Ontario portion of James Bay, including the drainage system for the Moose River, the largest river at the base of the Bay.

MANITOBA HYDROELECTRIC DAMS

Cree communities to the west are also affected by huge dam projects. In the early 1970s, a series of dams and power stations were built on the Nelson and Churchill Rivers systems, generating large amounts of power. Between them, these systems drain one of the largest watersheds in North America, causing silting, which chokes the reservoirs, causes widespread mercury contamination, and destroys wildlife.

A major consequence has been widespread economic dislocation. At Moose Lake, for instance, two-thirds of the communities' land base was flooded and hundreds of people were moved into a housing project. Before the dam, at least 75 percent of the food and the majority of the income came from the land. Today this is impossible, and most people must buy food at stores, often at extraordinarily high prices. Elsewhere, suicide epidemics, substance abuse, and other psychological problems flooded the communities.

Today, hydroelectric dams exist or are planned throughout much of northern Canada, from Newfoundland to British Columbia. In all cases Native people face relocation and devastation of their ecosystems, while most of the electrical power is designated for U.S. markets.

NUCLEAR-WASTE CONTAMINATION: HANFORD

Nuclear waste remained the largest obstacle for a peaceful atom, and Native peoples are again central to the discussion. The Hanford nuclear reservation is well within the treaty area of the Yakima Indian nation on the Columbia River. A significant portion of the 570 square miles of land contained in the nuclear site is contaminated. Approximately twenty different indigenous peoples reside in this area. In August, 1973, over 115,000 gallons of high-level liquid radioactive waste seeped into the ground from a leaking storage tank. The waste contained cesium 137, strontium 90, and plutonium, one of the most toxic substances known to humans. At least 400,000 gallons of radioactive material have been reported as having leaked at the Hanford reservation.

Soil at Hanford is so contaminated that officials classified and removed it as high-level radioactive waste. The U.S. Department of Energy changed its definition of nuclear waste so that plutonium contamination can increase almost indefinitely at the site. Airborne dust released from smokestacks cannot be and is not contained by the site boundaries.

The U.S. government recently solicited every Indian tribe within U.S. borders to host a possible nuclear waste storage facility. Officials entice tribes with "no strings attached" grants of hundreds of thousands of dollars. The federal office of Nuclear Waste Negotiation states its mission as finding "a state or Indian Tribe willing to host a repository or monitored retrievable storage facility for nuclear waste..." Some reservations accept the offers, but most reject these overtures.

These experiences are not unique. They are indicative of the conflict between industrial society and the land and the people who live on it. Hundreds of additional examples might be cited from Hopi, Zuni, Acoma, Isleta, Crow, Northern Cheyenne and other native nations in the United States, and from the Cree, Metis, Athabasca, and other native peoples of Canada.

As resource extraction plans and energy megaprojects are proposed for indigenous lands, there is neither consideration for the significance of these economic systems nor their value for the future. A direct consequence is that environmentally destructive development programs often ensue, foreclosing the opportunity to continue the intergenerational economic practice that had been continuous. For many indigenous peoples, the reality is that, as sociologist Ivan Illich has noted, "the practice of development is in fact a war on subsistence."[6]

CONCLUSION

Indigenous peoples remain on the front lines of the North American struggle to protect our environment. We understand clearly that our lives, and those of our future generations, are totally dependent on our ability to continue resisting colonialism and industrialization in our lands and to rebuild our communities.

In the United States, examples abound of indigenous peoples' successful battles to defend their homelands. These are just a few where Native peoples successfully resisted the destruction of their land and lives:

- The Northern Cheyenne, who now face a series of coal developments, have successfully defeated and opposed coal strip-mining on their reservation for almost three decades.
- The Gwich'in people, and a coalition of environmental groups, successfully averted the opening of the Arctic National Wildlife Refuge in the fall of 1991.

- The White Earth Anishinabeg from northern Minnesota successfully impeded the siting of a nuclear waste repository on the reservation.
- The Cree of the Moose River, near James Bay, have thus far stopped the proposal for twelve dams in the river basin.

1992 marked the 500th anniversary of the invasion. Many Native people would say that the invasion is the problem. The denial of the invasion and its consequences, for example, the destruction of more species in the past 100 years than the period since the ice age, indicates the lack of awareness of industrial societies for the land and its people.

Indigenous nations continue their resistance to industrialization and to a way of life not based on *mino bimaatisiiwin.* Many Native peoples believe that it is in the interest of all North Americans to join this struggle. Indigenous nations also see themselves in a critical dilemma in the entire industrial framework: providing sources for resource extraction while simultaneously maintaining living models of sustainable development, even after 500 years. We hope that North American peoples will recognize these indigenous models and values as the basis for future discussions of sustaining life on the earth.

An Ojibway prayer:

Megwetch
Wabunong
Zhawanong
Ningabi' anong
Giwedinong
Megwetch.

Notes

1. See *World Economic Survey, 1992: Current Trends and Policies in the World Economy* (New York: United Nations, 1992): 11.
2. Robert Venables, "The Cost of Columbus: Was There a Holocaust?" *Northeast Indian Quarterly* (Fall 1990).
3. Ward Churchill and Winona LaDuke, "The Political Economy of Radioactive Colonialism," in M. Annette Jaimes, ed., *The State of Native America: Genocide, Colonization, and Resistance* (Boston: South End Press, 1992), 241.
4. See OECD Nuclear Energy Joint Report with the International Atomic Energy Agency, 1988; and Winona LaDuke, "Native America: The Economics of Radioactive Colonialism," *Review of Radical Political Economics* 25 (Fall 1985).
5. Andre Picard, "James Bay II," *The Amicus Journal* (Fall, 1990).
6. Radio interview with Ivan Illich by David Cayley, Canadian Broadcasting Company (Fall, 1990).

10

Blue-Collar Women and Toxic-Waste Protests

The Process of Politicization

Celene Krauss

THE PAST TWO decades have witnessed the politicization of everyday life, as seen in the emergence of grass-roots protests around the world. These protests, often led by women, involve such issues as housing, community control, and the defense of the environment, in addition to the many grass-roots organizing efforts more immediately associated with the women's movement, such as wife battering and rape crisis centers.

Toxic wastes have become a major focus of grass-roots activism in recent years.[1] In particular, many blue-collar women have organized around these issues, responding to the threat they pose to the family.[2] This article examines the process by which blue-collar women become politicized through grass-roots protest activities around issues of toxic wastes and the implications of their politicization for social policy and social change. Calling themselves the "new environmental movement," these grass-roots protesters bear little resemblance to the more middle-class activists who are involved in national environmental organizations. This movement attracts a diverse constituency that cuts across race and class lines, including working-class housewives and secretaries, rural African American farmers, Navaho Indians, and low-income urban residents. According to the Citizens Clearinghouse for Hazardous Wastes (CCHW), an organization created to meet the needs of grass-roots activists, 80 percent of the leaders of grass-roots protests are blue-collar women.

The grass-roots activities of these women challenge both traditional assumptions of the policy-making process and left perspectives on social change. The traditional

---

This article originally appeared in the proceedings of the Second Annual Conference, Institute for Women's Policy Research, Washington, D.C., May 1991. Reprinted by permission of the author.

policy-making process focuses on dominant interest groups and institutions, providing few avenues for participation of blue-collar women. Their protests are too often trivialized, ignored, and viewed as self-interested and parochial, failing to go beyond a single-issue focus. This view of community grass-roots protests holds true not only for policy makers but for many analysts of movements for progressive social change as well.[3] As Bookman and Morgen have noted in their study of working-class women's struggles, women's political activism in general and working-class political life at the community level in particular remain "peripheral to the historical record...where there is a tendency to privilege male political activity and labor activism."[4]

In contrast to these views, I propose another way of understanding the process by which blue-collar women become involved in grass-roots community protest around toxic wastes. It is true that blue-collar women become involved around particular single-issue protests: They fight to close down toxic-waste dump sites, prevent the siting of hazardous-waste incinerators, influence chemical companies' production processes and waste disposal, and push for recycling projects. However, when these women become involved in the policy-making process, they inevitably encounter a world of power that excludes them. As they reflect on their political experience, they develop a broader analysis of government power as shaped by gender and class. This transformation in consciousness informs the direction of protest, their larger political critique, and the resources of struggle. This more elusive process, generally hidden from view, reveals an important dimension of toxic-waste activism. The understanding of inequities of power is not a priori. In the process of becoming activists, blue-collar women who believed in the system dramatically shift perspective in their understanding of political life. They recognize the failure of the system as a whole to act on their behalf and their own disenfranchisement from the policy-making process. They have to find ways to expose this and make the system democratic. Their activism challenges the traditional policy-making process in order to make the system work.

THEORETICAL PERSPECTIVES

Two bodies of writing help to explain the significance of grass-roots protest activities. The first is feminism;[5] the second is the new social historiography.[6] Both perspectives locate power in ordinary people by positing a theory of politics and political change that begins with the "particular," the everyday world of experience, and the centrality of consciousness as an agency of change. Feminists and new social historiographers shift the focus for social change from the macroanalytic level of social structure, political elites, and public policy to daily life. They create an alternative to top-down views of politics and political change that mask a complex relationship between everyday life and the larger structures of public power, rendering the lives, experiences, and the potential for political agency of most people invisible.

Feminist theorists have challenged a dominant ideology that separated the public world of policy and power from the private world of everyday experience. This ideology relegated the lives and concerns of most women to the private, nonpolitical arena. "Men took on the state," and deemed that significant, "and left the care of civil society to women."[7] This division in theory and practice constituted the political silencing of women as women's lives, traditionally particularistic and local, were ignored as "men moved onto other, more important matters."[8] Most significant, the dominant ideology made women participants in their own subordination. Inasmuch as women internalized the separation between the private and the public, they came to interpret their own lives through an ideology not of their making and were constrained from becoming agents of social change.

The women's movement took as its central task the reconceptualization of the political itself, critiquing the ideology that separated the public and the private and constructing a new definition of the political, locating it in the everyday world of ordinary people rather than in the world of public policy and public power. In so doing, feminists also helped make visible a new dimension of social change. If oppression is viewed as not merely externally imposed but also internally reproduced through ideology, then social change necessarily involves a subjective process as well. For feminists, the development of critical consciousness through the critical reflection on the everyday world of experience is the important subjective dimension of social change. Central to feminist theory and practice is the notion of consciousness-raising, the reinterpretation of the individual, private experience of oppression as a public, political issue. Critical consciousness makes it possible for women to uncover the complex and often elusive links between their private troubles and public power. It is a way of making visible the social relations that shape the everyday world, of coming to understand that the particular and personal always "exist within a determinate relationship with the wider social organization."[9] As women unmask the ideology that hides the relationship between the public and private, they may begin to understand the particulars of their own lives in new ways and to act politically. Discovering the link between the personal and political in everyday life seems like a small change, but it may be a "decisive factor determining its course and outcome."[10] Forging the link between the personal and the political empowers women to construct new definitions of reality and to believe in their own ability to act. The feminist emphasis on the development of critical consciousness offers important insights into the analysis of grass-roots protest around toxic-waste issues: Until women have made the connections between particular health problems in their own lives and the larger world of public policies and power that cause them, they cannot act politically.

The second body of work that helps to explain grass-roots protest is social historiography as advanced by writers such as E. P. Thompson, George Rudé, and Sheila Rowbotham. These writers help to illuminate a poorly understood dimension of working-class struggles, the ways they grow out of working-people's culture, traditions and community:

> classes are historical configurations and...the roots of radicalism lie in a people's inherited culture....The argument is not that a people's traditions are inherently radical, but that those traditions provide the categories out of which radicalism will develop, if it will develop at all....Popular culture was not a drag on progress...it was the spur to action.[11]

George Rudé notes that it is often difficult to understand the ideologies of struggle because ordinary working people appropriate and reshape popular cultural meanings, such as traditional beliefs about family and community. In the process they undergo a "sea change" that may be difficult for observers to understand.[12]

This analysis is critical to understanding the struggles of blue-collar women rooted in community. Since the protest of blue-collar women around toxic-waste issues is framed in the traditions of motherhood, family, and community, these struggles often appear particularistic and parochial. Progressive social analysts have viewed an emphasis on localism and community with suspicion because they see local culture as inherently conservative. Terry Haywoode argues in her study of working class feminism that not only policy analysts and left political theorists but also feminists themselves have sometimes undervalued the significance of these community struggles by seeing them as voluntary organizations and missing the process of empowerment for working-class women who shape new organizations.[13] Rather than being a drag on progress, these everyday concerns about family and community are the levers that set in motion the political protests of working-class people, who fight to protect the violation of deeply held traditions and values. These traditions and values in turn shape the language and oppositional consciousness that emerge.

THE EXPERIENCE OF TOXIC-WASTE PROTESTERS: THE PROCESS OF POLITICIZATION

What I examine is the process by which blue-collar women, who were never involved in the political system, become politicized as they are forced to confront inequities of power in the political system, trivialization of their involvement on the part of middle-class, male governmental officials, and growing conflicts with their spouses as a result of their involvement in protest activities. The analysis is based on interviews, conference presentations of activists, and writings of women involved with the Citizen's Clearinghouse for Hazardous Waste, a national organization created by Lois Gibbs as a resource for community groups struggling with toxic-waste issues around the United States. Gibbs is best known for her successful campaign to relocate families in Love Canal, New York, when residents discovered that they lived adjacent to a hazardous toxic dump site that endangered their health and their lives.[14]

Family and the Inequities of Public Power

Complicating women's involvement in toxic-waste issues is the exclusion of most people from the policy-making process. Generally, government makes toxic-waste policies without the knowledge of community residents. People may not know that they live near a toxic dump, or they assume that it is regulated by government.

Toxic-waste dump sites have been located primarily in blue-collar and people of color communities; hence, it is not surprising that blue-collar women have played a significant role in this movement. Because women are traditionally responsible for the health of their children, they are the most likely to make the link between toxic waste and their children's ill health. In communities around the United States, women are discovering numerous toxics-related health hazards: (e.g., multiple miscarriages, birth defects, cancer deaths, and neurological symptoms). Their initial response to this discovery is to contact their government, because they assume that government will protect the health and welfare of their children. Said Gibbs:

> I grew up in a blue-collar community, it was very patriotic, into democracy...I believed in government...I believed that if you had a complaint, you went to the right person in government. If there was a way to solve the problem, they would be glad to do it.[15]

But Gibbs and others faced an indifferent government. At Love Canal, local governmental officials argued that toxic-waste pollution was insignificant, the equivalent of smoking just three cigarettes a day; in South Brunswick, New Jersey, local officials argued that living with pollution was "the price of a better way of life."[16] State officials often withheld information from residents because "they didn't want to panic the public," further undermining confidence in government. For example, at Stringfellow, California, where 800,000 gallons of toxic-waste chemicals pumped into the community flowed directly behind the elementary school and into the playground, children played in puddles contaminated by toxic wastes, yet officials withheld information for fear of panicking the public.

Faced with government's indifference to a dangerous toxic-waste situation and its impact on their families, many women became activists, turning to protest activities in order to protect their families. At Love Canal, for example, women vandalized a construction site, burned effigies of the governor, and were arrested during a baby-carriage blockade. Women justified their protests on the grounds that they were making the system do what it's supposed to do.

Women become involved in these issues because they are concerned about the health effects on their families. As they become involved, they must ultimately confront more than the issue of toxic wastes; they must confront a government indifferent to their needs. In doing so, they learn about a world of power usually hidden from them:

> All our lives we are taught to believe certain things about ourselves as women, about democracy and justice, and about people in positions of authority. Once we become involved with toxic waste problems, we need to confront some of our old beliefs

> and change the way we view things....We take on government and polluters....We are up against the largest corporations in the United States. They have lots of money to lobby, pay off, bribe, cajole and influence. They threaten us. Yet we challenge them with the only things we have—people and the truth. We learn that our government is not out to protect our rights. To protect our families we are now forced to picket, protest and shout.[17]

As a first step, blue-collar women are forced to examine their own assumptions about political power. As they become involved, they see the contradiction between a government that claims to act on behalf of the public interest, a government that holds the family sacrosanct, and the actual policies and actions that government pursues. They see the ways in which government policy often favors the wishes of powerful business interests over the health and welfare of children and their families. The result is a more critical political stance that informs the militancy of their activism.[18]

Class, Gender and Resources of Power

As blue-collar women become involved in toxic-waste issues, they also come into conflict with a public world where policymakers are traditionally white, male and middle-class. As Zeff *et al.* has noted:

> Seventy to eighty per cent of local leaders are women. They are women leaders in a community run by men. Because of this, many of the obstacles that these women face as leaders stem from the conflicts between their traditional female role in the community and their new role as leader: conflicts with male officials and authorities who have not yet adjusted to these persistent, vocal head-strong women challenging the system....Women are frequently ignored by male politicians, male government officials and male corporate spokesmen....[19]

Many women who become involved in grass-roots activism do not consider themselves activists and have rarely been involved in the political process. Initially they may be intimidated by the idea of entering the public arena. Male officials further exacerbate this intimidation by ignoring women, criticizing them for being overemotional, and especially by delegitimizing their authority by labeling them "hysterical housewives," a label widely used regardless of the professional status of the woman. As a source of empowerment, blue-collar women have appropriated the criticisms and used them against their critics. Notes Cora Tucker, who organized a toxic-waste fight in rural Halifax, Virginia:

> When they first called me a hysterical housewife I used to get very upset and go home and cry....I've learned that's a tactic men use to keep us in our place. So when they started the stuff on toxic waste....I went back and a guy gets up and says, "We have a whole room full of hysterical housewives today, so men we need to get

> prepared." I said, "You're exactly right. We're hysterical and when it comes to matters of life and death, especially mine, I get hysterical." And I said, "If men don't get hysterical, there's something wrong with them." From then on, they stopped calling us hysterical housewives....[20]

The language of "hysterical housewives," "emotional women," and "mothers" has become a language of critique and empowerment that exposes the limits of the public arena to address the importance of family, health, and community.

The traditional role of mother as protector of the family can empower blue-collar activists. Their view of this role provides the motivation for women to take risks in defense of their families. Lois Gibbs, for example, has written of her insecurity when she first went to her neighbors' homes to conduct a health survey of her community at Love Canal. When she approached the first house and no one answered, she ran home relieved. Then she thought about her son and his seizures, gathered up her courage and tried again and again. On another level, these women have learned to derive power from their emotionality, a quality valued in the private sphere of family and motherhood but scorned in the public arena:

> What's really so bad about showing your feelings? Emotions and intellect are not conflicting traits. In fact, emotions may well be the quality that make women so effective in this movement....They help us speak the truth.[21]

Grass-roots activists look to their experiences as organizers of family life as a further source of empowerment. Lois Gibbs notes that women organized at Love Canal by constantly analyzing how they would handle a situation in the family and then translating that analysis into political action. For example, says Gibbs,

> If our child wanted a pair of jeans, who would they go to? Well they might go to their father since their father had the money—that meant that we should go to Governor Carey.[22]

Gibbs notes that the Citizen's Clearinghouse for Hazardous Wastes conducts women's organizing conferences, which help them learn to translate their skills as family organizers into the political arena.

Finally, blue-collar women recognize the power they wield in bringing moral issues to the public, exposing the contradiction between a society that purports to value motherhood and family, yet creates social policies that undermine these values. As Gibbs notes:

> We bring the authority of mother—who can condemn mothers?—it is a tool we have. Our crying brings the moral issues to the table. And when the public sees our children it brings a concrete, moral dimension to our experience—they are not an abstract statistic.[23]

The private world of family is a resource used as they struggle to protect their families in the public sphere. In the process of protest, blue-collar women come to see the ways in which class and gender shape inequities of traditional political power and the contradiction between the values of family and community espoused by those in power and the actual policies and actions that government pursues. This analysis informs their consciousness and the direction of their action and becomes a resource to empower them.

Gender Conflicts in the Family

As blue-collar women become activists, they are also forced to confront power relations and highly traditional gender roles characteristic of the blue-collar family. They begin to see the ways in which relationships within the patriarchal family mirror larger power relations within the society.

> Women's involvement in grassroots activism may change their views about the world and their relations with their husbands. Some husbands are actively supportive. Some take no stand: "Go ahead and do what you want. Just make sure you have dinner on the table and my shirts washed." Others forbid time away from the family.[24]

For many women, their experience contradicts an assumption in the family that both husband and wife are equally concerned with the well-being of children. As Gibbs notes,

> The husband in a blue-collar community is saying get your ass home and cook me dinner, it's either me or the issue, make your choice. The woman says: how can I make a choice, you're telling me choose between the health of my children and your fucking dinner, how do I deal with that?[25]

When women are asked to make such a choice—between their children and their husbands' needs—they see the ways in which the children are their concern, not their husband's. This leads to tremendous conflict and sometimes results in more equal power relations within the family, a direction that CCHW is trying to encourage by organizing family stress workshops. Generally, however, the families of activist women do not tolerate this stress well. Gibbs notes the very high divorce rate among activists and that, following protest activities, CCHW receives a higher number of reports of wife-battering. Thus, blue-collar women's involvement in grass-roots protests may disrupt the traditional organization of family life, revealing inequities in underlying power relationships. Ironically, the traditional family itself often becomes a casualty of this process.

IMPLICATIONS FOR PUBLIC POLICY AND SOCIAL CHANGE

For the most part, the formal policy-making process excludes the participation of those people most affected because they are seen as having only limited, particularlistic self-interests. In contrast, grass-roots protests concerning toxic wastes show us an informal process of resistance for people ordinarily excluded. In order to realize their political goals, they develop an oppositional or critical consciousness that informs the direction of their actions and challenges the power of traditional policymakers. When they see how the system excludes them, they go outside the formal policy-making process. When they cannot get the government to protect them from corporations, they use protest to force the political system to act on their behalf. Involvement in protest activities concerning toxic-waste issues ultimately leads many blue-collar women to a redefinition of politics and policy making from below, in which policies begin with the lives of those most directly affected. Their experience exposes the false assumption that the traditional policy-making process will be democratic and responsive to their needs. Rooted in the particular, they transcend the particular as they uncover and confront a world of political power shaped by gender and class.

The theoretical perspectives provided by feminism and the new social historiography help to make visible this process of politicization. As blue-collar women fight to protect their children and communities from hazards posed by toxic wastes, they reevaluate the everyday world of their experience. This leads to a critical examination of traditional assumptions about relationships in the family, community, and society. Traditional beliefs in the working-class culture about home, family, and women's roles serve a crucial function in this process. These beliefs provide the initial impetus for women's involvement in issues of toxic wastes and ultimately become a rich source of empowerment, as women appropriate and reshape traditional language and meanings into an ideology of resistance. Ultimately they develop a broader analysis of power as it is shaped by gender and class. The inequities of power between blue-collar women and middle-class male public officials exposes the contradiction between a government that claims to act on behalf of family and the actual policies and actions of that government, revealing traditional power relationships within the family. In these ways, blue-collar women come to reject the dominant ideology that separates the public and private arenas, rendering invisible and insignificant the private sphere of women's work. This enables them to act politically and, in some measure, challenge the social relationships of power. In so doing they implement an alternative view of policy making fundamentally more inclusive and democratic. Thus, their political activism has implications far beyond the visible single-issue policy concern of the toxic-waste dump site or the siting of a hazardous-waste incinerator. Fighting toxic wastes in defense of the family becomes the lever for the transformation of consciousness and the political action of blue-collar women.

NOTES

1. For important studies on toxic-waste activism, read Robert D. Bullard, *Dumping in Dixie: Race, Class, and Environmental Quality* (Boulder, Colo.: Westview Press, 1990); Phil Brown and Edwin J. Mikkelsen, *No Safe Place: Toxic Waste, Leukemia, and Community Action* (Berkeley, Calif.: Univ. of California Press, 1990); Michael R. Edelstein, *Contaminated Communities: The Social and Psychological Impacts of Residential Toxic Exposure* (Boulder, Colo.: Westview Press, 1988); and Lois Gibbs, *Love Canal: My Story* (Albany, N.Y.: State University Press of New York, 1982).
2. For more recent work on the role played by women in toxic waste protests, which has been published since this paper was first presented, read Sherry Cable, "Women's Social Movement Involvement: The Role of Structural Availability in Recruitment and Participation Processes," *Sociological Quarterly* 33 (1992); Cynthia Hamilton, "Women, Home, and Community," *Woman of Power* 20 (1991): 42-45; Celene Krauss, "Women and Toxic Waste Protests: Race, Class and Gender as Resources of Resistance," in Robert D. Bullard, ed., *Environmental Justice and Communities of Color* (Sierra Club Books, 1993); Mary Pardo, "Mexican American Women Grassroots Community Activists: 'Mothers of East Los Angeles,'" *Frontiers: A Journal of Women's Studies* 11 (1990): 1–7.
3. For critiques of grass-roots activism, read Robert Bellah, "Populism and Individualism," *Social Policy* (Fall 1985); and Hans Magnus Enzensberger, "A Critique of Political Economy," *New Left Review* (1974).
4. Sandra Morgen, "It's the Whole Power of the City Against Us! The Development of Political Consciousness in a Women's Health Care Coalition," in Ann Bookman and Sandra Morgen, eds., *Women and the Politics of Empowerment* (Philadelphia, Penn: Temple Univ. Press, 1988), 97. This extremely important book of studies shows the integration of class, race, and gender through grass-roots struggles.
5. Important feminist analyses that illuminate the transformation of consciousness through grass-roots activism include Martha A. Ackelsberg, "Communities, Resistance, and Women's Activism: Some Implications for a Democratic Polity," in *Women and the Politics of Empowerment*, ed. Ann Bookman and Sandra Morgen, (Philadelphia: Temple Univ. Press, 1988): 53-76; Patricia Hill Collins, *Black Feminist Thought: Knowledge, Consciousness, and the Politics of Empowerment* (Boston: Unwin Hyman, 1990); Nancy Hartsock, "Feminist Theory and the Development of Revolutionary Strategy," in *Capitalist Patriarchy and the Case for Socialist Feminism*, ed. Zillah R. Eisenstein, (New York: Monthly Review Press, 1979): 56-81; Sheila Rowbotham, *Women's Consciousness, Man's World* (New York: Penguin Books, 1973); and Dorothy Smith, *The Everyday World as Problematic: A Feminist Sociology* (Boston: Northeastern Univ. Press, 1987).
6. Important social histories focusing on the formation of class consciousness through concrete struggles include Herbert Gutman, *Work, Culture and Society in Industrializing America* (New York: Vintage Books, 1977); Sheila Rowbotham, *Women, Resistance and Revolution* (New York: Vintage Books, 1974); George Rudé, *Ideology and Popular Protest* (New York: Pantheon Books, 1980); and E. P. Thompson, *The Making of the English Working Class* (New York: Vintage Books, 1966).
7. Manuel Castells, *The City and the Grassroots* (Berkeley, Calif.: Univer. of California Press, 1983), 68.

8. Jean Bethke Elshtain, *Public Man, Private Woman* (Princeton: Princeton Univ. Press, 1981), 303.
9. Ibid., 305.
10. Thelma McCormack, "Toward a Nonsexist Perspective on Social and Political Change," in Marcia Millman and Rosabeth Moss Kanter, eds., *Another Voice: Feminist Perspectives on Social Life and Social Science* (Garden City, N.Y.: Basic Books, 1975), 9.
11. Jeff Lustig, "Contested Terrain: Community and Social Class," *democracy* 102 (April 1981).
12. George Rudé, *Ideology and Popular Protest* (New York: Pantheon, 1980).
13. Terry Haywoode, "Working Class Feminism: Creating a Politics of Community, Connection, and Concern,"(Ph. D. diss., The Graduate School and University Center of the City University of New York,1990).
14. Many of the quotes in this analysis are drawn from an interview with Lois Gibbs, director of the Citizen's Clearinghouse for Hazardous Waste (May, 1989 Cherry Hill, N.J.), and presentations of community activists at a Women in Toxics Organizing Conference, November, 1987, which were published by the Citizens Clearinghouse for Hazardous Wastes, 1989, in a pamphlet called *Empowering Ourselves: Women and Toxics Organizing* ed. Robbin Lee Zeff, Marsha Love, and Karen Stults.
15. Zeff et al., *Empowering Ourselves.*
16. Ibid.
17. Ibid., 31.
18. For a more in-depth analysis of the development of political consciousness, read Celene Krauss, "The Elusive Process of Citizen Activism," *Social Policy* 14 (1983): 50–55; and Celene Krauss, "Community Struggles and the Shaping of Democratic Consciousness," *Sociological Forum* 4 (1989): 227–238.
19. Zeff et al., *Empowering Ourselves,* 25.
20. Ibid., 4.
21. Gibbs interview.
22. Ibid.
23. Ibid.
24. Ibid., 33.
25. Ibid.

11

Acknowledging the Past, Confronting the Present

Environmental Justice in the 1990s

Richard Moore and Louis Head

THE 1980s SAW a quiet yet steady resurgence in the movement for social justice in the United States in which the term "environment" took on a new meaning both for poor communities and for those who historically had been known as "environmentalists." Grass-roots activists, organizers, and organizations working in communities of color began to link public-health issues, such as lead-based paint in housing projects; occupational safety concerns, such as pesticide poisoning of farmworkers; or the siting of waste disposal operations or dangerous industries in communities of color to a broader understanding of the relationship of development, its environmental consequences, and social justice. By the end of the decade, the recognition of such relationships and the increase of grass-roots efforts organizing to address such issues had resulted in a new unity of Asian, Pacific Islander, Native, Latino, and African Americans, the like of which had never been seen.

This growing unity has both strengthened the work of hundreds of grass-roots organizations and provided previously isolated communities with the capacity to win important victories. It has made it possible for communities of color in the United States to establish or reaffirm ties with grass-roots leaders and organizations abroad. It has also made possible meaningful partnerships between the white environmental movement and those who historically have struggled for social, racial, and economic justice.

This essay examines some major efforts of the environmental justice movement in the 1990s in its struggle for social, racial, and economic equality, and important dynamics characterizing the relationships between grass-roots organizations composed primarily of people of color and what has historically been defined as the "environmental movement." It draws primarily on the experiences of the Southwest Network for Environmental and Economic Justice and some of its affiliate organizations in playing their part building what has come to be defined as the environmental justice movement.

The year 1990 was a political bellwether for social justice and environmental activism in the United States. The Gulf Coast Tenants Organization and the SouthWest Organizing Project initiated open letters to the "Group of Ten" organizations calling for equitable distribution of resources and for representation of people of color on the boards and staffs of the major environmental players.[1] The newly formed Southwest Network for Environmental and Economic Justice called on environmentalists working in communities of people of color to provide resources to and accept direction from those affected communities and workers. Similar challenges from abroad were delivered during that same period by indigenous peoples' organizations calling on international conservation groups to respect sovereignty rights in the South and, likewise, work with indigenous people rather than against their interests.

Change has occurred, though perhaps not as much within the traditional environmental movement as some might have expected at the time the letters were sent. Several of the major organizations have taken positive though limited steps towards diversification. Some have begun to engage in partnerships with communities of people of color. More significant changes have occurred within two organizations outside the Group of Ten—the National Toxics Campaign (NTC) and to a lesser extent, Greenpeace—both of which had previously shown a greater commitment to working with people of color. Nearly one-half of the NTC board of directors, for example, is now composed of people of color with concrete experience in fighting environmental racism and injustice. Another group, the Environmental Careers Organization (ECO), has placed college-student interns with grass-roots groups and progressive think tanks such as the Panos Institute in Washington, D.C.; ECO had previously catered exclusively to the needs of large environmental groups and corporations.

Perhaps the single most visible difference during the 1990s is that people of color have begun to build their own movement, the viability of which was clearly demonstrated at the 1991 First National People of Color Environmental Leadership Summit in Washington, D.C. Rooted in previous and continuing struggles for civil and human rights, this movement defines the environment as where people live, work, and play and as integrally linked with issues of social and racial injustice that confront us daily.

While a range of perspectives exists within the movement regarding the strategic relationship of whites and people of color, most organizations have worked to develop tactical alliances with both white environmentalists and other white activists. In several cases whites have played an active role in building environmental justice organizations. However, the first priority of the movement has been to build unity among people of color so that they may determine their own needs and develop their own leaders, perspectives, and political agendas. "We speak for ourselves" has become the watchword of the movement.

This movement is exemplified by the Southwest Network for Environmental and Economic Justice. Founded in 1990 at a southwestern regional gathering of activists of color convened by the SouthWest Organizing Project, the Network has brought together over fifty grass-roots and indigenous organizations from Texas, Oklahoma, New Mexico, Colorado, Arizona, Nevada, and California and has emerged as a significant regional and national force in the struggle for environmental and economic justice. The Network exists both to increase the capacity of local organizations and communities and to make it possible for them to influence policy at all governmental levels. The Network has been successful, in part, because of the strength of some of its organizations.[2] But more so the Network's success results from its bringing communities and grass-roots organizations together based on what unites them in practice: common negative experiences with corporate, military, and government polluters and related economic impacts, coupled with an understanding that much more can be won working together with others than by attempting to fight alone.

The Southwest Network's wide range of activities have benefited local communities, strengthened the capacity of local organizations, and affected environmental policy at the national level. The Network Border Justice Campaign is addressing the severe public health crisis in Mexico-U.S. border communities resulting from the unbridled growth of multinational corporate activities in the region; it is developing a binational collaboration between U.S. and Mexican grass-roots organizations for the first time in twenty years. The Network High Tech Campaign directly confronts both the microelectronics industry and its backers within the U.S. government to cease the poisoning of high-tech chip production workers and surrounding communities, and to put a rein on the incessant capital flight and accompanying community disruption associated with the industry. The Sovereignty/Dumping on Native Lands Campaign raises awareness among people of color and others regarding the toxic assault against Indian people by waste disposal companies and other multinational corporations. These Network programs involve communities and workers directly affected by the problems being addressed.

TAKING ON THE EPA

The most far-reaching and successful programmatic effort conducted by the Network thus far has been its Environmental Protection Agency Accountability Campaign. The EPA Campaign was initiated in July 1991, when the Network submitted an open letter to the EPA that documented more than a decade's lack of enforcement of environmental regulations in communities of color in the region.[3] The Network spoke to cases where EPA had known of life-threatening problems and had failed to act, such as in South Dallas, where EPA had backed off shutting down lead smelters under industry pressure, despite the daily poisoning thousands of African

American and Latino American residents. In the predominantly Mexican community of Kettleman City, California, the EPA had approved a major expansion of a ChemWaste landfill without even holding a hearing for affected residents to voice their concerns. These, among many examples, including those from other regions of the United States, showed the EPA's negligence in dealing with issues important to communities of color. The Network targeted both the national EPA at the national level and specifically EPA Regions VI and IX, which include most of the southwestern United States. In both EPA regions, the front-line groups involved in the campaign were Network affiliates from communities towards which the EPA had shown a lack of accountability.

The Network chose to confront the EPA publicly because it had rarely if ever opened its doors to people of color or their organizations in the past. Demonstrations and press conferences in Dallas, San Francisco, and Albuquerque, New Mexico, accompanied the submission of the letter. When the EPA refused to respond, another round of demonstrations was held, including the occupation of the San Francisco EPA Region IX office. While EPA continued to give little public response to these events, top-level EPA officials prepared a lengthy confidential plan to use the national media to paint a positive image of the EPA's record on environmental equity. The memo termed environmental racism to be "one of the most politically explosive environmental issues yet to emerge."[4] Action needed to be taken, it was claimed, before resentment against the EPA reached a flashpoint at which "grass-roots groups finally succeed in persuading more mainstream (sic) groups to take ill-advised actions."[5] When the Network leaked the memo to the public via Congressman Henry Waxman (D-CA), the EPA began to seriously discuss the Network's concerns with the organization and other environmental justice leaders.

The basic thrust of the campaign was to force the EPA to enforce environmental regulations on an equal basis in communities of people of color and to promote the development of policies that would empower them to prevent the siting of dangerous and polluting industries in their communities. The process of forcing the EPA to the negotiating table with the Network was exciting in terms of organizing; several Network affiliates involved in the campaign had been attempting individually for years to get the EPA to act on their concerns. They began to see that regional and national collective direct action produced results.

The ensuing results of the campaign amazed even its own grass-roots protagonists. In communities from Texas to California, the EPA acted—under the Network's direction—to conduct health studies, democratize the decision-making process regarding siting of hazardous waste facilities, force the closure of dangerous industries, open up needed field services, and cease discriminatory practices in the relocation of residents of Superfund communities. In 1992 alone, EPA regional administrators in Dallas and San Francisco visited dozens of affected communities throughout the Southwest. The campaign was successful in promoting national meetings among the top leadership of the EPA, environmental justice activists, and academicians throughout the country. Finally, the campaign played a major part in forcing the EPA—which for years ranked forty-sixth out of the largest fifty government agencies in the hiring and promotion of people of color—to mount an equity program to diversify its staff and programmatically address issues of environmental racism.[6] While such initial internal moves left much to be desired, they represented a significant first step towards greater accountability by the EPA.

In the fall of 1992, then EPA Chief Administrator William Reilly met with the leadership of the Southwest Network in Albuquerque to discuss Mexico-U.S. border concerns, Native American sovereignty, and a host of issues relating to EPA practices in communities of people of color. The Network had begun its campaign literally locked out of EPA offices throughout the region. By the time Reilly left office, the Network had gained the agency's respect.[7]

Perhaps most important, the challenge to the EPA served to show those in doubt that grass-roots organizations of poor people could win against a powerful U.S. government entity by working together. Rather than relying on leadership from white environmental organizations, people of color developed their own agenda, program, objectives, strategies, and tactics in carrying out the campaign. They used independent, confrontational direct action in order to force the EPA to the bargaining table. In so doing, the Network attracted new grassroots affiliate organizations from throughout the region, thus increasing its own depth and strength.

DEVELOPING ALLIANCES

At the same time, however, the Network actively sought new types of relationships with historically white organizations. Shortly after its founding, the Network initiated a dialogue with several national and southwestern organizations that had previously shown a willingness to work with communities of people of color. In June, 1990, the Network Coordinating Council submitted a letter to nine organizations, including Greenpeace, the National Toxics Campaign, and the Citizen's Clearinghouse for Hazardous Wastes.[8] The Network expressed a clear appreciation for the efforts of these organizations to support grass-roots struggles in communities of color. However, the Network pointed out that these groups remained controlled by whites. If long-term meaningful partnerships were to be established between them and the environmental-justice movement that they claimed to support, they would have to share power and decision making with the grass-roots. As with the Group of Ten, the Network asked the progressive organizations to advocate *with* rather than *for*, the growing movement and communities of people of color.

The significant changes that have taken place at the National Toxics Campaign* are the result of both the development of leadership within the environmental justice movement and of an intense struggle that took place within the NTC. In late 1990

* The National Toxics Campaign dissolved in April 1993. Its Environmental Justice Program, however, continues to operate.

that organization requested and received multiyear funding commitments from several prominent progressive foundations to support a proposed effort to address environmental racism through a yet undefined effort within communities of people of color. While the Network and other environmental justice organizations welcomed such an initiative, it had been conceived primarily by white environmentalists working with funders accustomed to controlling the provision of resources for organizing efforts of, by, and for people of color. If NTC was to conduct such an effort, those directly affected had to be the primary decision makers in its development and implementation.

Based on the dialogue initiated in June, 1990, NTC had already begun to step up its placement of people of color on its board of directors. Those in place demanded and obtained greater numbers of their peers on the board and formed their own caucus in order to strategize over how to move the diversification process and program forward within NTC. After months of struggle and discussions, the NTC Environmental Justice Project (EJP), a partnership of people of color and white environmentalists, was born.

The EJP developed a comprehensive eighteen-month program that put aspiring grass-roots organizers through a month of intensive training and then placed them back in their own organizations and communities throughout the country. They contributed significantly to local and regional struggles. While it is still too early to evaluate fully the long-term success of the EJP, it represents a model partnership that has provided a means for the movement to combine its own human resources with outside support to develop and enhance leadership within the movement.

The success of the NTC project has been followed by others with similar dynamics. One important example, also involving the Southwest Network, has been the Campaign for Responsible Technology (CRT). In this case, Network affiliate organizations in Phoenix, Arizona, Oakland, California, Albuquerque, New Mexico, and Austin, Texas have combined efforts with the Silicon Valley Toxics Coalition and others to confront the poisoning of workers and communities by the microelectronics industry, which has fled to the Southwest in search of lower wage scales, tax breaks, and weak enforcement practices by state regulatory agencies in the region.

As with the NTC experience, the development of the CRT has involved a process of building trust and mutual respect between two camps. On one hand, there are those directly affected by the practices of high-tech industries. On the other are those who historically have had access to financial resources and traditional political channels but who have little history of working with or accepting leadership from people of color. An important initial victory for the CRT was to lobby the U.S. Congress in 1992 to divert $10 million of Department of Defense appropriations toward research to develop safe technologies for microelectronics production. The effort is currently organizing affected communities and high-tech workers to challenge and change the practices of corporations in each community and to influence national industrial policy that is placing the highest emphasis on high-tech development and production.

RESOURCING THE MOVEMENT

The NTC and CRT examples are clear victories in the struggle to democratize the control and use of resources. They show that the impact of funding and technical assistance goes far beyond increasing local capacity for grass-roots organizations. They make it more possible for the environmental and economic justice movement as a whole to assert its independence, and for our organizations to participate as equal players in any alliances. Such independence and equality across the negotiating table are preconditions for true partnerships and collaborative efforts to take place.

However, the funding provided to the National Toxics Campaign and the Campaign for Responsible Technology pales in comparison to that being offered to Group of Ten organizations for environmental justice or equity activities, let alone that allocated to the same organizations for projects that address conservation or other concerns long considered the exclusive domain of more affluent sectors of our society. The failure to provide resources to efforts of, by, and for people of color demonstrates the degree to which institutional racism pervades the traditional environmental movement as much as any other sphere of society. Such racism may or may not be intentional. What matters is that it is systemic and results in a skewed distribution of resources.

Most national nonprofit advocacy organizations espousing progressive agendas were molded into their present form during the Reagan-Bush era of the 1980s. Denied direct access to two administrations over a twelve-year period, they focused on influencing policy primarily through the legislative arena and electoral politics.

The major environmental organizations were no exception. Having come of age in the 1960s and '70s, despite their lack of direct access to government during the 1980s, many garnered financial support from major donors, public and private foundations and dues-paying members.[9] This support both reflected the interests of those who gave it and was given on the premise that policy is best developed at the national level in direct consort with elected officials and government agencies. The base for political action would arise as needed from middle- and upper middle-class sectors frightened by the utter extremes of the Reagan-Bush administration but accustomed to participating in electoral politics and other established traditional political channels.

The question of resource allocation, and its relationship to how policy is affected, was central to the challenge issued from the grass-roots to the environmental movement in 1990. Despite a decade of Reaganism, during the 1980s local groups had blossomed to fight against powerful corporate and government interests that were poisoning workplaces and communities. In spite of growing successes, such activity organized by poor and often sick people went unassisted by most of the environmental movement and by funders. While the major white organizations obtained access to millions of dollars and battalions of technical and legal experts, social-justice leaders and organizers were left to imagine what could be accomplished with significant monetary and technical support. Instead, a trickle-down approach resulted in a paltry amount of support reaching those on the front lines, usually from small foundations and intermediary organizations that, whether by intent or not, served as movement brokerage houses.

Grass-roots leaders and organizers engaging in dialogue and partnerships with environmental organizations have not waited for any flood of support based on the challenge to the Group of Ten. There has been an understandable hesitation to accept assistance from organizations that base their power on the support of affluent whites and traditional lines of political access. Instead, the power of the movement is based on its independence, its level of grass-roots democracy, and skills inherited from the past. Its greatest strength is that its leaders come from communities directly affected by environmental and economic injustice. As demonstrated by the EPA campaign and many other success stories, with virtually no outside funding and little technical assistance, people have fought and won victories that are leading to improved conditions in communities and policy changes at the national level.

A GRASSROOTS PERSPECTIVE

Many of today's grass-roots organizations were born in the early 1980s, primarily in response to the worsening conditions and the political atmosphere that were characterized by the ascendancy of the Reagan administration. Many had their roots in the 1960s, a period during which government commitments to fight poverty and institutionalized racism were won through grass-roots struggle. Leaders of these past struggles saw their efforts usurped by federal poverty programs and a myriad of emerging nonprofit social-service and policy organizations. The same sectors weakened the racial and social-justice movement of the '60s by taking organizers, activists, and leaders away from their communities and into social service and advocacy positions. With the many gains of the 1960s institutionalized in the form of federal programs and no accountability to the grass-roots, all it took was six months of the Reagan administration to wipe out most of what had been won in areas such as affordable housing, food and nutrition, affirmative action, education and health care.

Those who worked to reinvigorate grass-roots organizing in the 1980s did so with these lessons in mind. In contrast to the national liberal establishment, the grass-roots movement defined political participation based on those methods that had worked historically: direct action and democratic participation from below. A basic supposition was that accepted lines of political participation through electoral politics were essentially closed. The drive for accountability on the part of government structures—and our own leaders—to those affected communities became a strength of the emerging movement. This was in sharp contrast to liberal strategies and perspectives. In the words of SouthWest Organizing Project (SWOP) cofounder and director Jeanne Gauna, "when SWOP came together (1980), nobody was talking about community control, as if that were a backward notion. We were saying no, that the only way that you can really change things in this country is if the leadership and participation comes from affected communities."[10]

In spite of the many impressive gains made by grass-roots environmental and economic-justice organizations over the past decade, there has been limited acceptance of the contributions of the movement beyond occasional lip service. Evidence of this was clearly seen following the 1992 general elections, when organizations such as the Southwest Network had to force their way into the government transition process. Notwithstanding previously stated commitments from several Group of Ten organizations to work in partnership with the grass-roots, the major organizations—which were involved with the Clinton campaign and then the transition process immediately following the election—had to be sharply reminded by the Network and others of their commitments before they endorsed meaningful participation by the environmental justice movement in the transition process.[11]

Similar evidence of the need for continued struggle has emerged in the behavior of both foundations and the major environmental organizations in developing and implementing programs to address environmental injustice. In response to the call for environmental justice, a significant amount of foundation resources have been allocated for internal diversification or programmatic efforts to address environmental racism undertaken by several Group of Ten organizations. Yet directly affected communities have had to fight hard, with limited success, to participate fully in the development and implementation of such efforts meant to benefit them. Racial diversity is currently in vogue in corporate white America. The question to the environmental movement is whether diversification will mean the placement of people of color into token front-office positions, or a true restructuring in which those most directly affected by environmental degradation assume positions of power and equality so that environmental activism is democratized and can better merge itself with the struggle for social, racial, and economic justice. In so doing, the major environmental organizations indeed run the risk of alienating many of their wealthier supporters.

In spite of the roadblocks, working relations are taking shape between what have historically been divergent sectors. A tremendous amount of progress has been made in leveling the playing field between them. Such relationships have usually happened between more affluent whites and people of color, and have often suffered from all the contradictions that emerge from such relationships. The major barrier to multiracial and multicultural alliances, especially those inclusive of whites, is institutional racism. We have tried to show, based on concrete experience, that institutional racism has manifested itself with respect to how resources are controlled and who controls them, and how policy is defined and changed. These concerns were first raised in the letters to the environmental movement in 1990; the contradictions continue to be played out in the various dialogues, discussions, and partnerships that have taken place since that time between grass-roots and environmental organizations. The degree to which the barriers are honestly confronted and addressed will determine whether we can develop strong, productive and long-lived multiracial alliances.

NOTES

1. Those organizations that received letters from both groups included the National Wildlife Federation, Friends of the Earth, Wilderness Society, the National Audubon Society, the Sierra Club, the Sierra Club Legal Defense and the Education Fund, the Natural Resources Defense Council, and the Environmental Defense Fund. See Gulf Coast Tenants Organization and SouthWest Organizing Project letters, January and March, 1990, respectively. See also "Environmental Groups Told They Are Racists in Hiring," *New York Times*, 1 February 1990.
2. For example, the SouthWest Organizing Project was able to serve as the lead organization for the Network during its first two years. Other more established groups, such as the Southwest Public Workers Union, Citizens Alert Native American Program, Concerned Citizens of South Central Los Angeles, and Tucsonians for a Clean Environment, were able to give much-needed time and resources to the Network as well.
3. See letter to William K. Reilly, July 1991, from the Southwest Network for Environmental and Economic Justice.
4. See memorandum from EPA Associate Administrator Lewis Crampton to Chief of Staff Gordon L. Binder, 15 November 1991.
5. Ibid.
6. U.S. Equal Employment Opportunity Commission, Annual Report (Washington, D.C., 1989).
7. A much more comprehensive history and analysis of the Network EPA Campaign by Elizabeth Martinez will be available in late 1993.
8. See letter to Greenpeace, June 1990 from the Southwest Network for Environmental and Economic Justice.
9. Some groups, such as the National Wildlife Federation, accepted significant support from known corporate polluters, a major issue raised by the SouthWest Organizing Project in its letter to the Group of Ten.
10. Interview with Community Cable Channel 27, Albuquerque, October 1991.
11. See Memorandum Regarding Clinton Administration Transition Team Chronology, Southwest Network for Environmental and Economic Justice, December 1992.

12

Building on Our Past, Planning for Our Future

Communities of Color and the Quest for Environmental Justice

Vernice D. Miller

OVER THE LAST DECADE, communities of color across the United States have become aware of a new menace threatening our communities: the siting of environmentally hazardous facilities or substances in the midst of where we live. We want to stop the poisoning of our people and of our land. By linking environmental issues to social, racial, and economic justice, we have created a new and dynamic social-justice movement. The environmental-justice movement seeks to change forever the material conditions of people of color in the United States, as well as those whose lives are affected by U.S. policies throughout the world. Ultimately what we seek is a fundamental transformation of society where racial and economic justice prevail.

In 1987, the United Church of Christ Commission for Racial Justice published the landmark study *Toxic Wastes and Race*, which documented and quantified the phenomenon of "environmental racism." The study's findings "suggest the existence of clear patterns which show that communities with greater minority percentages of the population are more likely to be the sites of commercial hazardous waste facilities. The possibility that these patterns resulted by chance is virtually impossible, strongly suggesting that some underlying factors, which are related to race, played a role in the location of commercial hazardous waste facilities. Therefore the Commission for Racial Justice concludes that, indeed, race has been a factor in the location of hazardous waste facilities in the United States."[1]

As a research assistant working on the study, I remember feeling that, finally, somebody understood and believed what we in West Harlem had been saying since

1968—that our community was being used as a dump site for the City of New York primarily because its inhabitants were African American and Latino American.

Since 1987, communities of color across the United States have broadened the context of their struggles for justice to include environmental justice. Many communities that had been engaged in these struggles for some time have redefined their battle, as environmental justice has come to symbolize every aspect of the discriminatory and unequal treatment that we have been experiencing all along.

In every state, some community is organizing and fighting back against the evils of racial discrimination, poverty, and economic and environmental exploitation. The reverberations from these communities have reached a fevered pitch as the environmental-justice movement advances nationally.

THE FIRST NATIONAL PEOPLE OF COLOR ENVIRONMENTAL LEADERSHIP SUMMIT

In October 1991, under the sponsorship of the United Church of Christ Commission for Racial Justice, the First National People of Color Environmental Leadership Summit was held in Washington, D.C. According to a paper produced for the Summit, "environmental inequities cannot be reduced solely to class—the economic ability of people to 'vote with their feet' and escape polluted environments. Race interpenetrates class in the United States and is often a more potent predictor of which communities get dumped on and which ones are spared. There is clear evidence that institutional barriers severely limit access to clean environments. Despite the many attempts made by government to level the playing field, all communities are still not equal."[2]

As described by Reverend Ben Chavis in the preface to the proceedings of the Summit, "the Leadership Summit is not an independent 'event' but a significant and pivotal step in the crucial process whereby people of color are organizing themselves and their communities for self-determination and self-empowerment around the central issues of environmental justice. It is living testimony that no longer shall we allow others to define our peoples' future. The very survival of all communities is at stake."[3]

The Summit was a momentous event in the life of the environmental-justice movement. Standing together during the first two days of the Summit, we who had been selected as delegates were reinforced by the multicolored faces of three hundred people who each understood and shared a common commitment to the preservation of all our communities. During the opening session you could see a sea of bobbing heads as the delegates eagerly surveyed the large ballroom where we were all gathered. You could sense what we were all thinking: We're not alone anymore!

For three and a half days we as delegates, later joined by a group of three hundred or so participants and observers, struggled through many difficult moments. We had a very tense session with leaders from mainstream environmental organizations, and many emotional accounts of the real conditions faced by different communities. The conference produced several exercises in democratic process and collective decision making, as well as cross-cultural and group dynamics.

STRATEGIC PLANNING

During and after the Summit we devised several strategies to continue the struggle. We decided that those strategies would include four initiatives: local, state-wide, regional, and national. We also made a commitment to support the initiatives of communities other than our own wherever possible, including communities in the Third World.

The Summit conducted over twenty-three regional, strategy, and issue workshops. In these workshops we discussed and debated a wide range of issues and requirements for effective strategies. In addition to five regional workshops, others included grass-roots empowerment, law, health, labor, religion, education and youth, media, culture, intergroup cooperation, sustainable development and energy, capacity building, environment and the military, land rights and sovereignty, urban environment, impacting environmental decisionmaking, environmental health, occupational health and safety, and international issues. We tried to cover every aspect of the oppressive environmental conditions that most often threaten our communities.

OVERALL OBJECTIVES

The delegates at the Leadership Summit took great pains to draft and ratify the "Principles of Environmental Justice" and affirm the "Call to Action," since both of these documents articulate exactly what we mean when we say we want to transform society. Principles 1 and 5 perhaps best encapsulate the spirit of this movement: "Environmental justice affirms the sacredness of Mother Earth, ecological unity and the interdependence of all species, and the right to be free from ecological destruction," and "Environmental justice affirms the fundamental right to political, economic, cultural and environmental self-determination of all peoples."[4]

As a result of the Summit, and efforts prior to the Summit, several objectives have been set forth for the movement as a whole regarding issues of environmental equity. "[E]nvironmental equity falls into three major categories: *Procedural Equity,* [which] deals with the 'fairness' question, and the extent that governing rules, regulations, and evaluation criteria are applied uniformly across the board. *Geographic Equity,* [which] deals with the idea that some neighborhoods, communities, and regions are disproportionately burdened by hazardous waste. *Social Equity,* [which] deals with the role of race, class, and other cultural factors in environmental decision-making."[5]

The Summit workshops, and significant regional and national follow-up to the Summit objectives, have led this movement to address various equity questions through the following initiatives:

1) to influence and redefine federal legislation and policy-making regarding hazardous-waste production, disposal, and clean-up, and waste-facilities siting;
2) to challenge and eradicate unequal environmental protection by the EPA and other federal agencies, such as the Department of Energy, the U.S. Department of Defense, and the Nuclear Regulatory Commission;
3) to work with academics of color to conduct research that supports local community struggles;
4) to gain control of local political processes, and establish community control of local decision-making processes;
5) to elect or appoint "true" community representatives to decision-making bodies; and, most important,
6) to represent ourselves and speak for ourselves at the tables of power and authority and policy-making.

PLAN OF ACTION

Because the environmental-justice movement is so diverse and broad-based, activists pursue several different avenues of attack simultaneously. Local communities are continuing their front-line struggles against environmental racism in its many forms, including closer monitoring of and involvement in local land-use and zoning processes. We are also participating in local legislative processes to influence environmental policy-making as it directly affects our communities.

Based on our continued grass-roots organizing, we are starting to (or in some cases continuing to) coalesce regionally. This development enables us to address common obstacles by bringing our resources and people together in a more focused attack. Regional organizing also strengthens our ability to respond to the EPA with a united-front. In the Southwest this strategy has proven particularly effective under the auspices of the Southwest Network for Environmental and Economic Justice.

Many of the local strategies, vis-à-vis legislation and policy-making, are also being replicated on the state level, as state agencies continue to site waste facilities in communities of color. The united-front strategy is also being employed within states to coalesce local communities to affect state wide policy-making.

Many exciting developments are occurring nationally, as the environmental-justice movement broadens its focus to address the overarching political and legal framework that undergirds the phenomenon of environmental racism. A group of grass-roots activists, academics, national civil-rights organizations, tribal leaders, and environmental policy experts, all people of color, have been meeting with EPA officials and staff on a regular basis to advise them on policy vis-à-vis environmental-justice issues. Subsequently the EPA has formed a national environmental-equity interagency working group, and has initiated environmental equity projects in three EPA regions across the country. While this is a step forward for EPA, it is, nevertheless, a woefully inadequate response to the catastrophic environmental conditions that many communities of color face.

At this writing the reauthorization of the Resource Recovery Conservation Act (1992) and the Superfund Reauthorization Act (1993) have both had sustained involvement from representatives of the environmental-justice movement, and from members of the Congressional Black Caucus and their staff, in the redrafting of these two significant pieces of federal legislation to more accurately reflect the legislative needs of grass-roots communities. Also in 1992, then-Senator Albert Gore and Congressman John Lewis, in association with Reverend Ben Chavis, introduced the Environmental Justice Act. This legislation is a good start, but grass-roots activists believe that it still misses the mark and is not the type of comprehensive legislation that will help us put an end to the dumping of hazardous wastes and the siting of waste facilities in our communities and on our lands. During the 1992 presidential transition process, another historic step was taken as Dr. Robert Bullard was chosen to represent the environmental-justice community in the environmental-transition cluster, the first time that a grass-roots movement has ever had a representative in such a process.

Many academics of color are supporting the establishment of "communiversities" at historically African American colleges and universities. The intent of the communiversity is to link academic institutions to grass-roots communities and provide much-needed research and technical assistance to bolster local community struggles. This framework establishes that research must be linked to local community needs and provides a mechanism for accountability of the university to the affected communities. Dr. Beverly Wright is creating a communiversity at Xavier University in New Orleans. Another avenue, vigorously pursued, is encouraging college and graduate students of color to pursue careers in environmental fields, in order to develop the capacity for technical expertise within our own communities.

There is a growing movement among lawyers of all races to develop the legal theories and strategies to win cases in the courts that challenge the legality of environmental racism. Many examples abound of legal organizations joining forces with local communities and community-based organizations to file lawsuits challenging the practice of siting waste facilities in our communities and the negative impacts that many of these facilities produce once they are operational. These efforts would include actions by the Mothers of East Los Angeles and the NAACP Legal Defense and Education Fund; the Latino American community of Kettleman City, California, and California Rural Legal Services; West Harlem Environmental Action and the Natural Resources Defense Council; the residents of Wallace, Louisiana, and the Sierra Club Legal Defense and Education Fund; and the Save the Audubon Ballroom Coalition of New York City and the Center for Constitutional Rights. In 1992, Native Americans for a Clean Environment won a ten-year legal battle against Sequoyah Fuels Corporation (a verdict that is on appeal as of this writing). Another promising legal development is the establishment of the Environmental Justice Project of the Lawyer's Committee For Civil Rights Under Law in Washington, D.C., which seeks to provide legal and technical assistance to grass-roots communities.

INTERNATIONAL DEVELOPMENTS

Perhaps the most significant evolution in the environmental- justice movement is the linkages formed with our counterparts in Third-World nations. The United Nations Conference on Environment and Development (UNCED, also known as the Earth Summit) was a catalyst for many significant developments. In March 1992, the Highlander Center, the People's Alliance, and the New York City Coalition for Environmental Justice hosted the People's Forum in New York City to coincide with the fourth Preparatory Committee Meeting for UNCED, which was held at the United Nations during the entire month of March. The purpose of the People's Forum was to bring together grass-roots activists from all over the United States and their counterparts from the Third World to share our common experiences and develop strategies that could affect the negotiations at UNCED. This included a tense meeting with Ambassador William Ryan, chief U.S. representative to UNCED, regarding the participation of grass-roots activists on the official U.S. delegation to the Earth Summit. We suggested that there should be several grass-roots representatives on the U.S. delegation, and Ambassador Ryan responded that perhaps we could have one. As it turned out, there were three African American NGO observers on the official U.S. delegation at the Earth Summit, all of whom were supportive of the environmental-justice movement. At the behest of the environmental-justice movement, this was the first time that people of color outside of government were ever included as part of an official U.S. delegation to a United Nations sponsored conference. Don Edwards, then executive director of the Panos Institute USA, was designated by the environmental-justice movement as our representative on the U.S. delegation.

At the local level, another component of the effort to build the network internationally was the creation of a series of Toxic Tours, designed to take the international delegates, as well as those from other parts of the United States, into underdeveloped and environmentally degraded communities in New York City. We took bus loads of delegates to neighborhoods in Brooklyn, West Harlem, and the South Bronx. These tours enabled grass-roots activists in the United States to accomplish something that language and cultural barriers had too frequently prevented: a demonstration that poverty and underdevelopment exist even in the United States and that race, class, gender, and power are central forces of our oppression.

As one of the organizers of the Toxic Tours and the People's Forum, and as someone who actively participated in the fourth Preparatory Committee Meeting from start to finish, I can honestly report that these two events were seminal in the effect that they had on all who were fortunate enough to participate.

Several representatives of the environmental-justice movement attended the Earth Summit in Rio de Janeiro. Our delegation to the Earth Summit, lead by Dana Alston (then of the Panos Institute USA) included fifteen activists from across the United States representing the breadth and diversity of our movement. The Highlander Center also brought a multiracial delegation of grass-roots environmental activists to the Earth Summit. We participated in many meetings at the Conference covering a wide range of issues, including sovereignty and indigenous peoples, population and development, toxic wastes, and the NGO treaty process, in which activists from all around the world signed treaties with each other regarding all the issues being negotiated by government leaders at the official UNCED meeting.

One of our most significant meetings was with representatives from the Third World Network to discuss how we could improve communication between us after the Earth Summit concluded and how we could begin to develop transnational strategies to respond to our common problems. The first major effort in establishing this global network was a 1992 meeting hosted by the Third World Network in Malaysia, focusing on toxic waste, in which representatives of the environmental-justice movement participated. As reflected in the Principles of Environmental Justice, our movement recognizes that the issues cannot be constrained by the territorial borders of the United States. Environmental justice is a global movement that seeks to preserve and protect global ecosystems as well as all the people that populate this planet. We are convinced that we cannot resolve our problems unless they are addressed in a global context through global solutions.

CONCLUSION

As an organization that I cofounded and work with locally, West Harlem Environmental Action has grown and developed in step with the environmental-justice movement as a whole. We began as an ad hoc community coalition in 1986, and now we are a freestanding community-based organization that provides information, technical assistance and expertise, resources, and community-organizing skills to our community and to other communities like ours around New York City who are addressing the multiple environmental plagues that have beset our communities. We have greatly benefited from the wisdom of many people with whom we have been privileged to work in this movement. We have drawn strength from the trials of our brothers and sisters all over this country. What we have learned from them can not be quantified. My comrades and I feel honored to be a part of this movement, a movement that I am certain will ultimately bring about justice for all.

NOTES

1. United Church of Christ Commission for Racial Justice, *Toxic Wastes and Race in the United States* (1987): 23.

2. Dana Alston and Robert Bullard, *People of Color and the Struggle for Environmental Justice,* a position paper on behalf of the National Planning Committee for the First National People of Color Environmental Leadership Summit, October 1991.
3. Charles Lee, ed., *Proceedings, First National People of Color Environmental Leadership Summit* (New York: United Church of Christ Commission for Racial Justice, Dec. 1992).
4. *Principles of Environmental Justice,* The First National People of Color Environmental Leadership Summit, 27 October 1991 (for the full text of the principles, see Apendix).
5. Ibid.

13

Unequal Protection

The Racial Divide in Environmental Law

Marianne Lavelle and Marcia A. Coyle

ACCORDING TO A *National Law Journal, (NLJ)* investigation, the federal government, in its cleanup of hazardous sites and its pursuit of polluters, favors white communities over minority communities under environmental laws meant to provide equal protection for all citizens.[1]

In a comprehensive analysis of every U.S. environmental lawsuit concluded in the past seven years, the *NLJ* found penalties against pollution-law violators in minority areas are lower than those imposed for violations in largely white areas. In an analysis of every residential toxic-waste site in the twelve-year-old Superfund program, the *NLJ* also discovered the government takes longer to address hazards in minority communities, and it accepts solutions less stringent than those recommended by the scientific community. This racial imbalance, the investigation found, often occurs whether the community is wealthy or poor.

Since 1982, activists with ties to both the civil-rights and environmental movements have been pressing their case for environmental justice before the U.S. Congress and the Environmental Protection Agency (EPA). Using an increasing body of scientific study, they have shown that minorities bear the brunt of the nation's pollution.

But the *National Law Journal*'s investigation for the first time scrutinized how the federal government's policies of dealing with polluters during the past decade have contributed to the racial imbalance.

The lifethreatening consequences of these policies are visible in the day-to-day struggles of minority communities throughout the country. These communities feel

---

This is a slightly revised version of an article that originally appeared in a special section of the *National Law Journal* (21 September 1992).

they are victims three times over—first by polluters, then by the government, and finally by the legal system.

African American families in South Chicago wonder whether the rampant disease among them springs from the fifty abandoned factory dumps that circle their public-housing project, and why the federal government won't help them. In Tacoma, Washington, where paper mills and other industrial polluters ruined the salmon streams and way of life of a Native American tribe, the government never included the tribe in assessing the pollution's impact on residents' health. And, nine years after an a Latino American neighborhood in Tucson, Arizona, poisoned by chemically tainted water, drew federal attention to its problems, nothing has been done to stop the migration of seepage as it creeps underground.

Emerging from communities like these across the country is the contour of a new civil-rights frontier, a movement against what activists charge is pervasive environmental racism. Whether pushing the edges of constitutional law or shaming the environmental establishment into opening up its own white-dominated boards and membership, this movement calls not for equity but for prevention and equal protection.

The following are key *National Law Journal* findings, gathered over an eight-month period and based on a computer-assisted analysis of census data, the civil court case docket of the EPA, and the agency's own record of performance at 1,177 Superfund toxic-waste sites.

- Penalties under hazardous-waste laws at sites having the greatest white population were about 500 percent higher than penalties at sites with the greatest minority population. Hazardous waste, meanwhile, is the type of pollution experts say is most concentrated in minority communities.
- For all the federal environmental laws aimed at protecting citizens from air, water, and waste pollution, penalties in white communities were 46 percent higher than in minority communities.
- Under the giant Superfund clean-up program, abandoned hazardous-waste sites in minority areas take 20 percent longer to be placed on the national priority action list than those in white areas.
- In more than one-half of the ten autonomous regions that administer EPA programs around the country, action on cleanup at Superfund sites begins from 12 to 42 percent later at minority sites than at white sites.
- At the minority sites, the EPA chooses "containment," the capping-or walling-off of a hazardous dump site, 7 percent more frequently than the clean-up method preferred under the law, permanent "treatment," which eliminates the waste or rids it of its toxins. At white sites, the EPA orders treatment 22 percent more often than containment.

A RACIST IMBALANCE

EPA lawyers, while declining to respond directly to the *National Law Journal's* analysis, say they carry out the law, case by case, on the basis of the science of, the size of, and the legal complications particular to each toxic-waste site or illegal pollution case.

"'Environmental equity' is serious business for this agency," says Scott Fulton, EPA deputy assistant administrator for enforcement. "We want to guarantee that no segment of society is bearing a disproportionate amount of the consequences of pollution."

But activists who have been working in communities inundated by waste say that the hundreds of seemingly race-neutral decisions in the science and politics of environmental enforcement have created a racist imbalance. Through neglect, not intent, they say minorities are stranded on isolated islands of pollution in the midst of the nation that produced the first, most sophisticated environmental-protection laws on earth.

"People say decisions are made based on risk assessment and science," says Professor Robert D. Bullard, a sociologist at the University of California-Riverside, who has been studying environmental racism for fourteen years. "The science may be present, but when it comes to implementation and policy, a lot of decisions appear to be based on the politics of what's appropriate for that community. And low-income and minority communities are not given the same priority, nor do they see the same speed at which something is perceived as a danger and a threat."

Many activists argue the result has been a less-safe environment for all citizens, as polluters' use of politically weak minority communities creates a gateway for disposal of wastes that will ultimately affect the larger environment. The lead particles that rise in West Dallas fall on Dallas, they point out, just as the chemical stew that starts near slums on the Mississippi pollutes the fishing source of the Gulf, and South Chicago dumps threaten the grand reservoir of the Midwest, Lake Michigan.

"In the metaphor of a rapidly sinking ship, we're all in the same boat and people of color are closest to the hole," says Deeohn Ferris, a former EPA official who is now environmental coordinator at the Lawyer's Committee for Civil Rights Under Law. That is why the most hopeful of environmental justice advocates believe that if they can force federal leaders to factor in race and poverty in making decisions, it could revolutionize and improve the law.

"This issue has the power to change the fundamental assumptions of environmental protection," says Charles Lee, director of the United Church of Christ's special project on toxic injustice, which conducted ground-breaking research on the issue in 1987. Indeed, one wry observation on how much pollution minority communities suffer was made privately by members of an EPA work group that for two years studied environmental equity. The success of their labors, they mused, was the best chance for achieving the Bush administration's more publicized goal, pollution prevention.

Others within the Bush administration viewed the issue with alarm. Said one administration official in a confidential memo early in 1992: "Long-simmering resentment in the minority and native American communities about environmental fairness could soon be one of the most politically explosive environmental issues yet to emerge."

A NEW MOVEMENT

The movement against environmental racism began to coalesce in 1982 with a church-led protest by Black residents against a toxic landfill in North Carolina that led to 500 arrests. Minority community leaders today in towns like Wallace, Louisiana, and Moss Point, Mississippi, have taken up the fight, standing firm against two of the most reviled of pollution threats, a hazardous-waste burner and a paper factory, that want to set up shop in their backyards. They hope to build upon attempts made since 1979 to use the law and the courts to mete out environmental justice, as ground-breaking and as difficult an effort as the first equal-protection cases that outlawed school segregation forty years ago.

"This is the cutting edge of a new civil-rights struggle," says Wade Henderson, director of the Washington office of the NAACP. "For our organization, it is a new and important area of activity."

It has been a difficult struggle, however, for communities that bear all of the other historical disadvantages of racism, such as lack of education and money. Consequently, some community organizers are aiming to create a new, strong civil-rights movement, one that will link the money, contacts, and legal expertise of the large national environmental groups with the grass-roots people who are tackling local problems.

But the nation's handful of mainstream green groups have been criticized roundly for their role in shaping the twenty-two year history of environmental law, a story of progress that nevertheless has left behind groups without a strong voice or scientific know-how.

Professor Richard J. Lazarus, of St. Louis's Washington University School of Law, uses the example of the lobbying frenzy around the Clean Air Act of 1990 as a process that has excluded some of the people, particular urban minorities, who suffer from toxic air pollution the most. Environmentalists are accustomed to criticism from political conservatives and those who favor greater industry autonomy, says Professor Lazarus. But the environmental-racism charge has troubled the conscience of a movement accustomed to thinking of itself as progressive.

"The fact is that all environmental statutes...pick winners and losers," he says. "They pick between problems, because there aren't enough resources to deal with all problems. And certain solutions redistribute risks, the most obvious example being that when you move a hazardous dump, one location is gaining and one is losing." Professor Lazarus argues that minority communities have been the biggest losers.

ASSESSING PENALTIES

The most striking imbalance between whites and minorities in the *NLJ's* analysis of the EPA's enforcement effort was a 506 percent disparity in fines under the Resource Conservation and Recovery Act (RCRA), the thirteen-year-old law that supposedly assures the safe handling and disposal of hazardous waste. The average fine in the areas with the greatest white population was $335,566, compared to $55,318 in the areas with the greatest minority population.

"This particular statistic is probably the most telling," says Arthur Wiley Ray, the head of environmental programs for Baltimore Gas & Electric Company, a former EPA lawyer who says he spent much energy during his government tenure urging the agency to heed the environmental-racism issue.

RCRA cases, he says, target active toxic-dump sites. And in the view of minority communities, "that's where the problem is," he says. "They don't put those dumps on Rodeo Drive; they put them across the tracks."

The landmark 1987 United Church of Christ study, *Toxic Waste and Race*, found that communities that had two or more active hazardous-waste plants or major landfills had three times as many minorities as communities without such facilities.

Minority communities also saw far lower fines than white areas in the handful of multi-law cases, twenty-eight of them, that the EPA has concluded from 1985-1992. In those, fines were 306 percent higher in white areas than in minority—$239,000 compared to $59,429.

Only in Superfund enforcement cases, lodged mainly against polluters who have been recalcitrant about cleaning up abandoned toxic-waste sites, did fines in minority areas come out higher than in white areas, by 9 percent. Minority communities saw lower average penalties in federal enforcement of the Clean Water Act, by 28 percent; the Clean Air Act, by 8 percent; and the Safe Drinking Water Act, by 15 percent.

The EPA says that many factors affect its assessments of penalties, such as the seriousness of an offense, the ability of a polluter to pay, the polluter's history and level of cooperation, and all the whims of judges and the legal system.

The EPA's Mr. Fulton calls penalties "an unreliable point of departure" for studying equity. He says the EPA is considering using some other benchmark of enforcement effectiveness to study EPA equity. For example, the agency may consider the number of inspections at a facility or the amount of time between the uncovering of a violation and the lodging of charges.

But Professor Bullard, the environmental-racism scholar, says penalties are a key component of deterrence. And, he argues, violators are driven to minority communities because penalties there are low enough to be discounted as a cost of doing business.

"What the companies are trading off is a minuscule part of the profit," he says. "What the residents living in impacted areas are trading off is their health. Right now, we have not seen a balance between economic development and people's health. Facilities operating in communities of color can operate with impunity."

One law-enforcement factor that observers say leads to inequity was apparent in the *National Law Journal*'s statistics: lack of resources. There have been few court cases at all—only sixty-five concluded in the seven years studied under the hazardous-waste law. Even under the Clean Air Act, with more cases than any other law, only fifty suits have been concluded annually, an average of one per state per year.

"It is clear that the environmental statutes promise a great deal," says Professor Lazarus. But everyone knows that these laws are not self-enforcing. Those who complain, who have greater access, who know how to tweak their Congress people to do something, are more likely to get the attention of very busy people. And the people with greater know-how are generally those with greater political and economic resources, who tend to be white."

In one area for which EPA has hard data, enforcement focuses more on white communities than on polluted communities. At the Department of Energy's Argonne National Laboratory researchers found that a greater percentage of the U.S. African American and Latino American populations live in areas where pollution levels are high enough to violate the standards of the Clean Air Act. But the population that has benefited from the 352 Clean Air Act cases in the last seven years, the *National Law Journal* found, is 78.7 percent white, 14.2 percent African American and 8.2 percent Latino American.

SUPERFUND DELAYS

Community activism gave birth to the most ambitious environmental program in the world, Superfund, and many believe that progress in this thirteen-year-old program still requires the political access and financial resources so scarce in minority communities.

The *National Law Journal's* investigation of the EPA's Superfund program shows that, for the sites with the most minorities, it took an average 5.6 years from the date a toxic dump was discovered to the date it arrived on Superfund, 20 percent longer than the 4.7 years it took for the sites with the highest white population.

EPA officials respond that the pace of action in the Superfund program depends upon how long it takes to study the hazards and to assess the risk to people at hazardous-waste sites. Urban sites may have a more complex mix of pollutants that therefore take longer to study. On the other hand, a rural site may be many miles wide and therefore, may take much longer to assess.

In 1992 Richard J. Guimond, then deputy assistant administrator of the EPA's office of solid waste and emergency response, which manages Superfund, said he could not draw conclusions from the *National Law Journal's* analysis. He said the EPA is attempting to study whether there is a disparate impact on minority communities in Superfund by comparing toxic sites that are similar in makeup.

"We realize in some cases we don't have all the information that would enable us to fully conclude whether there are inequities, as an artifact of the way things operate in society, what might be the reasons and the best way to deal with them," he says.

Latino Americans who live near a lead-smelter site in West Dallas are suing the EPA, charging environmental racism was the reason they could not gain Superfund status for their toxic sites. Similar complaints arise among African Americans who live near abandoned steel plants in Chicago and from a Latino American community near an Air Force plant in Tucson, Arizona, which saw no action or slow action on their problems in the Superfund program. All three of these communities complained of a high incidence of cancer, lupus, nerve damage, and birth defects; scant on money and expertise, they feel saddled with the burden of proving the link between disease and the toxics. "It's almost as if they have to convince the powers that be that these are problems, whereas other communities can use elected representatives, zoning boards, and commissioners to cut through that particular process," says Professor Bullard.

Once a site is placed on the Superfund list, the *National Law Journal*'s investigation shows that the pace of action speeds up for minority sites. By the time the comprehensive cleanup of a site begins, minority sites fall only 4 percent behind the white sites, 10.4 years compared to 9.9 years.

But the pattern is quite uneven across the country. In six of EPA's ten regional field offices across the country, where most Superfund decisions are made, the pace from the discovery of a site to the beginning of cleanup is from 8 to 42 percent faster at white sites than at minority sites. The greatest disparity is in midwestern Region 5, with the most sites (257), where the pace from discovery to cleanup was 13.8 years for minority sites compared to 9.7 years for white sites. In one area, midatlantic Region 3, the pace for minority and white sites is dead even.

In three regions, cleanup begins more quickly at minority sites than white sites: in the Deep South, Region 4, by 8 percent; in New York/New Jersey, Region 2, by 11 percent; and in the Pacific Northwest, Region 10, by 36 percent.

One indication of how successfully residents have lobbied for permanent and complete cleanup of Superfund waste is in the "remedial decisions" arrived at through negotiation by EPA, polluters, state authorities, and other interested parties. EPA categorizes these decisions each year as "treatment" or "containment," in response to Congress's order in 1986 to make treatment the preferred choice.

The more intensive treatment choice was chosen 22 percent more frequently than containment at the white sites; at minority sites, containment was chosen 7 percent more frequently.

POLITICAL CLOUT HELPS

The EPA says it is a simplification to judge its decisions at sites strictly by whether containment or treatment is chosen, although many studies by industry, environmental groups and the government itself have done so. Mr. Guimond says that EPA acts immediately at every site to remove unstable canisters and other materials that are considered an imminent threat to health.

In the Superfund decision-making realm, the EPA argues that its decisions are based on the science of particular sites, not on race. But in a program as massive and costly as Superfund, political clout certainly does help a community get solutions. Unfortunately, environmental-justice activists argue, white communities usually have been better able to gain this access than minority communities. A classic illustration is the difference in the treatment of two heavily polluted neighborhoods whose plights won the attention of Congress: an African American middle-class community in Texarkana, Texas, that was ignored for years, and a white trailer park in Globe, Arizona, that received immediate Superfund attention.

One scholar, Professor Paul Mohai of the University of Michigan School of Natural Resources, says that this difference stems from the classical effects of racism in U.S. society. Minorities continue to be underrepresented at every level of government and on the boards of polluting companies, he points out. And housing discrimination prohibits minorities from escaping their pollution problems, he says. Social-science studies have shown that minorities, especially African Americans, live in segregated enclaves in the United States, even as their level of income and education increases.

That's why activists like Deeohn Ferris of the Lawyers' Committee for Civil Rights Under Law are asking the EPA to begin to take into account disparate racial impact in addition to scientific analysis in making its decisions. This would be analogous to the Reagan-era directives that required federal agencies to consider the cost to industry with every decision.

Reverend Ben Chavis, former executive director of the United Church of Christ's Commission for Racial Justice and a founder of the environmental-justice movement, agrees that the EPA needs to rethink how it does business. "So much of the methodology of the last twelve years in environmental protection has been risk assessment and therefore risk management, and too little attention has been paid to equal enforcement of the law," he says. EPA officials, without concluding that racism results from their current methods, have begun to study how to do their job more equitably.

But Reverend Chavis says that, through litigation and organization, minority communities that are suffering the heavy costs of industrial pollution are not waiting for the EPA to do more studies. "'Cancer Alleys' serve as a reminder that the issue of the environment for us is an issue of life and death," he says. "There is a sense of urgency and wanting to ensure there will be no more 'Cancer Alleys', or Columbia, Mississippis, or South Side Chicagos.

"In each one of these areas," he says, "people are fighting back." Even in the worst situations, glimmers of hope emerge because people are uniting—across racial and socioeconomic lines—and the common demand is environmental justice.

NOTE

1. See *National Law Journal*, special section, September 21, 1992 for details.

14

Ecofeminism and Grass-roots Environmentalism in the United States

Barbara Epstein

ECOFEMINISM AROSE TO provide a framework for a feminist-inspired, women's environmental movement. Why has women's environmental activism taken different directions? What does this suggest about the relation between theory and activism? What does it suggest about what a feminist environmental strategy might be? As a background to speculating about these questions, I present a brief account of the history of ecofeminism and a brief survey of the role of women in the grass-roots environmental movement in the United States.

Ecofeminism appeared, in the late 1970s, as an intellectual and political trend at the intersection of the feminist, environmental, and peace movements. It flourished as a political tendency in connection with the antinuclear movement of the early 1980s, playing a part in both of that movement's phases, opposition to nuclear energy and to nuclear weapons. Early ecofeminist writings appeared sporadically through the 1970s. But ecofeminism as a political orientation took hold in the late 1970s, at a moment when many women who had spent years building a radical women's movement (or countercultural women's communities) wanted to join peace, environmental and other movements for broad social change and wanted to explore the relevance of a feminist perspective to global issues of militarism and environmental crisis. In the late 1980s and early 1990s ecofeminist writings have proliferated; much of it is theoretically nuanced and sophisticated. Ironically, the activist movement that provided the base for ecofeminism a decade ago has largely disappeared. As the environmental movement as a whole grows, women play an increasingly important role in it, especially in community-based environmental groups; but often such women do not identify themselves as feminists, let alone ecofeminists.

Emerging out of the radical feminist movement of the late 1960s and early 1970s, ecofeminism extended the critique of relations between the sexes by embarking upon an exploration of the implications of gender arrangements for the relations between humans and the environment. The term "ecofeminism" was first used by the French author Francoise d'Eaubonne in 1974, in her book *La Feminisme ou la Mort.* In 1980, a group of women who organized a conference entitled "Women and Life on Earth: Ecofeminism in the 1980s" adopted the term ecofeminism. The conference led to the first Women's Pentagon Action.[1] Ynestra King's role as conference coordinator, and the participation of others from the Institute for Social Ecology, helped further the development of ecofeminism from both political and theoretical perspectives.

In the late 1970s, a number of books attempted to join radical feminist and ecological concerns, including Susan Griffin's *Woman and Nature* and Mary Daly's *Gyn/Ecology.* In the 1980s many more works have appeared that identify themselves with ecofeminism: Carolyn Merchant's *Death of Nature,* Charlene Spretnak's collection, *Politics of Women's Spirituality,* and an issue of *Heresies* devoted to feminism and ecology. Margot Adler's account of witchcraft in the United States, *Drawing Down the Moon,* and Starhawk's *Dreaming the Dark: Magic, Sex and Politics,* both straddle the worlds of paganism, feminist spirituality, and ecofeminism.[2] In 1986, a conference entitled "Ecofeminist Perspectives" in Los Angeles attracted roughly fifteen hundred activists and scholars, suggesting a growing audience for this approach.

Ecofeminism argues that patriarchy, the domination of women by men, has been associated with the domination of nature. Men have justified their attempts to dominate nature by associating it with women, objectifying both women and nature by placing them in category of the "other." Patriarchy also involves a denial of human links with the natural world and of men's feminine side. Ecofeminists regard both the despoliation of the environment and violence and militarism as rooted in the culture of domination; they argue that both have become major threats to the human race. Patriarchy, ecofeminists argue, must be replaced with an egalitarian form of social organization in which men and women have equal power, and by a social ecology in which the natural environment is treated with respect and sustained rather than manipulated and destroyed. Ecofeminists also believe that capitalism is linked to domination and must be replaced by another social order; they envision small-scale economies and local, grass-roots democracy rather than the large-scale, state-directed societies and economies of existing socialist nations.

Ecofeminism has been strongly influenced by anarchism, especially by Murray Bookchin's countercultural politics and utopian vision. Like Bookchin, ecofeminists reject the Marxist tendency to privilege the economic realm over the cultural; they reject the Leninist concept of the revolutionary party, supporting instead a concept of nonviolent revolution that would dismantle rather than seize state power. But, like Bookchin's work, ecofeminism has roots in early Marxist theory, especially in the concept of alienation and vision of a socialist or communist society that would liberate human potentiality. Ecofeminism has also relied on the Frankfurt School and critical theory (e.g., Max Horkheimer's argument in *Eclipse of Reason* that social repression also requires the repression of nature, meaning both human nature and the natural environment)[3]. Perhaps the greatest strength of ecofeminism is that it has developed a theory that addresses both gender relations and the relationship between human society and the natural environment, showing how the structures of domination in these realms shape and reinforce one another. But ecofeminism has not gone very far towards addressing how these linked structures of domination might be dismantled. Calling for a movement that understands these connections is not the same thing as proposing a concrete strategy for addressing them.

Early writers in the ecofeminist tradition, such as Mary Daly and Susan Griffin, were closely associated with radical feminism, arguing that women should identify with nature against men and endorsing the development of a separate women's culture and a separatist strategy. But, as ecofeminism has developed, some who identify with it have distanced themselves from the radical feminist perspective. Ynestra King's approach, for instance, stands somewhere between radical feminism and traditional Marxism. She has criticized Marxism, and the socialist feminist tradition that looks to it, for excessive emphasis on the economic realm, for a tendency to subordinate questions of gender to those of class, and for a "rationalist severance of the woman/nature connection [in] advocating the integration of women into production, [its failure to] challenge the culture-versus-nature formulation itself."[4] But instead of aligning herself with radical feminism, King has criticized the radical feminist-socialist feminist split and has attempted to transcend it. This split, she argues, reflects the historical division between rationalism and romanticism, a manifestation of the nature vs. culture opposition in which women's oppression is rooted. "If the nature/culture antagonism is the primary contradiction of our time," King writes,

> it is also what weds feminism and ecology and makes woman the historic subject. Without an ecological perspective which asserts the interdependence of living things, feminism is disembodied.... Ecological feminism...is about connectedness and wholeness and the return of all that has been denigrated and denied to build this hierarchical civilization with its multiple systems of dominance. It is the potential voice of the denied, the ugly and the speechless—all those things called 'feminine.' It is no accident that the feminist movement rose again in the same decade as the ecological crisis. The implications of feminism extend to issues of the meaning, purpose and survival of life.[5]

Some strands of ecofeminism, closely associated with women's spirituality, have fostered a belief in a prehistoric matriarchal golden age. Many within the women's spirituality movement believe that, in the earliest human societies, goddesses were worshipped, and women held positions of power. These peaceful societies, in which people lived harmoniously with one another and with their neighbors, treated the natural environment with respect. According to this account, this stage of historical development was followed by patriarchy, which uprooted earlier earth-based goddess religions in favor of various monotheisms. These patriarchal societies became the bases for cultural imperialism, each of them worshipping a single, transcendent, male God and attempting to impose the same belief on other cultures. This account links patriarchy and monotheism with war and the domination of nature as well as that of women. It suggests that this stage of development may have been necessary as a spur to certain kinds of technological development but argues that it has outlived its usefulness and has become a threat to the human race and to the earth. This summary of world history suggests that survival rests on the ability of the human race to proceed to the next stage of development, which will be neither matriarchal nor patriarchal but based on equality between men and women and the abolition of social hierarchies generally. It suggests that such a society would honor many of the qualities traditionally associated with women—such as nurturance, intimacy, and sensual pleasure—and would devalue militarism, competition, and the love of power, which have been developed in a patriarchal context.

The idea of a golden age of matriarchy was presented as the dominant position in Charlene Spretnak's *The Politics of Women's Spirituality.*[6] It has also been encouraged by Merlin Stone's *When God Was a Woman*[7] and by Riane Eisler's *The Chalice and the Blade.*[8] Matriarchy here means a social system organized around matriliny and involving goddess worship in which women are accorded positions of power. These books point to evidence that there have been such societies, and that these societies were not destructive of their natural environments and lived in peace with their neighbors. The assertion that matriarchy was a stage in social development was associated with primitive communism going back to Friedrich Engels. In the twentieth century, this view has been opposed by the dominant trend in anthropology, on grounds that the patterns of goddess worship and matrilocality that evidently existed in many paleolithic societies were not necessarily associated with matriarchy defined as women's power over men. Many societies can be found that exhibit these qualities alongside female subordination. Furthermore, militarism, practices that destroy the natural environment, and hierarchical social structures can be found in societies in which goddess worship, matrilocality, and matriliny exist.

Janet Biehl, in *Rethinking Ecofeminist Politics*, attacks ecofeminism for devaluing rationality by blurring the distinction between human consciousness and the natural world, for offering instead a spirituality that regards nature as a living being. Biehl disagrees with many of the claims of ecofeminism, arguing that the myth of a golden age of gender equality and peace runs counter to anthropological evidence, and that the natural world is not a living being. She also argues that these beliefs have reactionary implications. A utopian view of early societies argues for a return to the past, while a view of the world as a living being celebrates the world as it is, including the societies that are a part of that world. Rationality, the elevation of human consciousness above nature, allows us to criticize the existing order. Finally, Biehl argues, as have many other feminists, that ecofeminism equates woman with nature, reinforcing sexist conceptions of gender.[9]

Biehl's reading of ecofeminism is too simple. Many ecofeminists would avoid making any claims for a prehistoric golden age, or claims that goddess worship necessarily corresponded with peaceful, egalitarian societies.[10] The ecofeminist view of nature as a living being is, for many, a way of emphasizing that the world around us is not merely an inert object of human manipulation but a realm with its own autonomous set of logic.[11] Because a conception of the world as a living entity implied acceptance of the status quo in premodern societies does not mean that such a conception always has the same meaning. In the late twentieth century, the concept of a living nature is a basis for protest against attempts to dominate nature. It is also a protest against the kind of rationality that separates thought from emotions, that celebrates thinking and denigrates feelings. Biehl's claim that ecofeminism equates woman with nature, reinforcing sexist conceptions of gender, is, like many of Biehl's charges, true of some ecofeminists and not of others. Women who came to ecofeminism through women's spirituality may believe that women have a special affinity with nature.

But others argue that a socially constructed connection between women and nature gives women a special vantage point from which to redefine what it means to be human by pointing to the connections between humanity and the natural world, connections that have been denied by male-centered rationalism. Carolyn Merchant examines what she describes as a socially constructed link between women and nature in the context of capitalism. "Both women and nature are exploited by men as part of the progressive liberation of humans from the constraints imposed by nature. The consequence is the alienation of women and men from each other and both from nature."[12] Ynestra King writes that all varieties of feminist theory have addressed the relationship between women and nature; the distinctive contribution of ecofeminism is to have resisted a dualistic understanding of this relationship:

> Ecofeminism takes from socialist feminism the idea that women have been historically positioned at the biological dividing line where the organic emerges into the social.... An ecological feminism calls for a dynamic, developmental theory of the person—male and female—who emerges out of nonhuman nature, where difference is neither reified nor ignored and the dialectical relationship between human and nonhuman nature is understood.[13]

Biehl sees ecofeminism's orientation toward spirituality as evidence that it is bankrupt. But the orientation toward spirituality gives ecofeminism much of its vitality—and also has been the basis for creating bonds between white women and women of color.[14]

The current upsurge in grass-roots environmentalism has very different sources. Antitoxics activity provides the focus for about five thousand groups around the United States; such groups have emerged in middle-class and working-class neighborhoods, among whites and people of color. Some grew out of community groups that opposed or addressed the effects of plant closures; others grew out of unions (most notably the Oil, Chemical and Atomic Workers), African American churches, or civil-rights groups. Some have been based on groups of neighbors organizing in response to dramatic local or regional threats to health, as in "Cancer Alley," Louisiana, and Love Canal, New York. The Highlander Educational and Research Center in Tennessee has in recent years devoted considerable attention to organizing around toxics. In the 1970s and '80s many of the civil-rights and other activist groups connected with Highlander began to turn to issues of toxics and health; as a result Highlander formed the Community Environmental Health Project, which addresses issues of race, class, the economy, and the environment. The antitoxics movement is racially mixed and includes groups from a range of class backgrounds. On the whole, however, it does not include the groups that have provided the basis for ecofeminism, the youth counterculture, and the university.

Environmental justice has emerged as an important issue within the grass-roots environmental movement. As a result, the movement has been a fertile arena for multiracial organizing and for debate about the relationship between race and class. Some groups representing communities of color argue that toxic dumping is primarily a racial issue, that the environmental-justice movement must be largely led by people of color. Other groups, especially those from all-white rural communities, experience toxic dumping as an issue of class. The relationship between race and environmental issues is prominent throughout the antitoxics movement. A People of Color Caucus has formed within the National Toxics Campaign. In October 1991, the First National People of Color Environmental Leadership Summit was held in Washington, D.C., largely involving representatives of antitoxics groups. The Environmental People's Forum, held in March 1992, and organized by a multiracial coalition of regional environmental justice groups, brought together 110 representatives of local groups, both white and people of color, to examine race and toxic dumping and relations between whites and people of color within the movement.

Though ecofeminism is not a significant presence in the antitoxics movement, women make up a large and growing part of the movement, and they bring to it a concern with gender equality and women's issues more generally. Women have a significant role in the leadership of the movement. Two national organization, the National Toxics Campaign and the Citizen's Clearinghouse for Hazardous Waste, provide resources for local groups and create connections among them. The latter is led by Lois Gibbs, who came into the antitoxics movement through her involvement in the struggle against toxics in Love Canal. On the local level, in both of these organizations, large numbers of women hold leadership positions, probably considerably more than do men.[15]

In spite of the large numbers of women in the environmental movement, and their increasingly prominent roles in leadership, women in the movement have not made issues of gender central to their political practice in the way that people of color have made issues of race. This is probably partly because the grass-roots environmental movement emphasizes the impact of toxics on particular geographical areas, highlighting race, not gender. It also seems likely that feminism has been a general influence rather than a driving political force in most of the communities in which antitoxics groups have emerged. Women who join the movement may be influenced by feminism, but they are not likely to define themselves primarily in relation to it. People of color, on the other hand, are likely to enter the antitoxics movement out of some prior involvement with movements concerned with racial issues. Even if they have no prior political experience, they usually identify themselves politically in relation to questions of race.

In the mid-1980s, ecofeminists put forward a vision of a movement made up largely if not entirely of women, certainly led by women, that would bridge issues of the environment and militarism, and would be infused with feminism. The current grass-roots environmental movement does not look like this, nor is it likely to in the near future; the only point of overlap is the significant role of women in environmental groups, and their large and probably growing role in the leadership of the movement.

Ecofeminism remains relevant to environmentalism. Ecofeminists rightly argued that environmental politics were likely to have a special attraction for women and that there is a connection between the definition of gender and the exploitation, and destruction, of the natural environment. Exploring this connection provides insight into the attraction of women to environmental politics. Exploring the connection between gender and the relationship between society and the natural environment also raises questions about what values govern our society and what values we want it to be governed by. Ecofeminism addresses the desire for a more sustaining relationship between society and nature, a desire better addressed by movements oriented toward spirituality than those oriented toward political action. But at this point the most pressing need of the environmental/antitoxics movement is not for a long-range vision or a theoretical framework within which to understand existing patterns of domination but a point of leverage, a way of focusing the widespread discontent over the degradation of the environment and the hazards posed by toxics, especially in communities of working-class people and people of color. The growing involvement of women in antitoxics and other environmental movements suggests that women may be especially receptive to, or willing to act upon, environmental issues. Perhaps we will see the rise of environmental movements in which women constitute a majority or at least the leading force—but which are not women's movements in the sense of restricting membership to women. Feminism would certainly play an important role in defining the objectives of such movements (e.g., they would no doubt include attention to environmental hazards to women's health) and in shaping their internal practice (in insisting upon internal democracy eliminate obstacles to women's full participation). Though the environmental/antitoxics movement is not oriented toward the theoretical issues addressed by ecofeminism, it is likely to become increasingly ecofeminist by combining environmental concerns and concerns about gender and domination in its political practice.

NOTES

1. The first Women's Pentagon Action took place in Washington, D.C., in 1980.
2. Susan Griffin, *Woman and Nature: The Roaring Inside Her* (New York: Harper & Row, 1978); Carolyn Merchant, *The Death of Nature: Women, Ecology and the Scientific Revolution: A Feminist Reappraisal of the Scientific Revolution* (New York: Harper & Row, 1980); Charlene Spretnak, ed., *The Politics of Women's Spirituality: Essays on the Rise of Spiritual Power Within the Women's Movement* (New York: Doubleday/Anchor, 1982), Mary Daly, *Gyn/Ecology: the Metaethics of Radical Feminism* (Boston: Beacon Press, 1978); *Heresies: a Feminist Journal of Art and Politics* (February 1983); Margot Adler, *Drawing Down the Moon: Witches, Druids, Goddess-Worshippers, and Other Pagans in America Today* (Boston: Beacon Press, 1979); Starhawk, *Dreaming the Dark: Magic, Sex and Politics* (Boston: Beacon Press, 1982).
3. Max Horkheimer, *The Eclipse of Reason* (New York: Oxford Univ. Press, 1947).
4. Ynestra King, "Feminism and the Revolt of Nature," *Heresies: a Feminist Journal of Art and Politics* (February 1983): 14.
5. Ibid., 15.
6. Spretnak, *The Politics of Women's Spirituality.*
7. Merlin Stone, *When God Was a Woman* (New York: Harcourt Brace Jovanovich, 1976).
8. Riane Eisler, *The Chalice and the Blade: Our History, Our Future* (San Francisco: Harper & Row, 1987).
9. Janet Biehl, *Rethinking Ecofeminist Politics* (Boston: South End Press, 1991).
10. Identifying ecofeminism with this perspective would exclude Carolyn Merchant and Ynestra King from the ranks of ecofeminists, while in fact both are regarded as leading ecofeminist writers.
11. See Donna Haraway, "Situated Knowledges: The Science Question in Feminism and the Privilege of Partial Perspective," in *Simians, Cyborgs and Women: The Reinvention of Nature* (New York: Routledge, 1991), 183-202.
12. Carolyn Merchant, "Ecofeminism and Feminist Theory," in Irene Diamond and Gloria Feman Orenstein, *Reweaving the World: The Emergence of Ecofeminism* (San Francisco, Sierra Club Books, 1990), 103.
13. Ynestra King, "Healing the Wounds: Feminism, Ecology, and the Nature/Culture Dualism," in Diamond and Orenstein, *Reweaving the World*, 116-117.
14. For instance, the ecofeminist WomanEarth Institute was quite successful in bringing together women of color and white women. The Institute held a conference in Amherst, in June 1988, at which there were equal numbers of women of color and white women.
15. On grass-roots environmentalism, especially the antitoxics and environmental-justice movements, see Harriet G. Rosenberg, "The Home Is the Workplace: Hazards, Stress and Pollutants in the Household," in Sedef Arat-Koc, Meg Luxton, and Harriet Rosenberg, *Through the Kitchen Window: The Politics of Home and Family* (Toronto: Garamond, 1990) 59-79; and Harriet Rosenberg, "'From Trash to Treasure': Housewife Activists and the Environmental Justice Movement," in Rayna Rapp and J. Schneider, eds., *Festschrift*

in Honour of Eric Wolf, forthcoming. I am also indebted to Florence Gardner of the University of California at Berkeley Department of Geography and the Highlander Center Environment and Democracy Campaign for her description of the current state of grassroots environmentalism.

15

The Effects of Occupational Injury, Illness, and Disease on the Health Status of Black Americans

A Review

Beverly Hendrix Wright
and Robert D. Bullard

THE UNITED STATES has made great strides in improving the health and longevity of its people. Although improved, the health status of some minority groups has shown a persistent and distressing disparity in important health indicators when compared to those of their white counterparts. In 1983, the life expectancy of whites reached a new high of 75.2 years, while African American life expectancy reached only 69.6 years. The present life expectancy of African Americans was reached by white Americans in the early 1950s, a lag of about thirty years.[1] Approximately 15 percent of the total U.S. population in the fifteen-or-under age category are African American. However, by the time they are sixty-four the relative proportion of the African American U.S. population decreases to 8 percent.[2]

To what do we attribute the cause of this disparity in life expectancy among African Americans as compared to whites? Although many factors are presumed to influence African American health status and life expectancy in the United States today, researchers consider environmental and occupational exposures as major sources of disease and illness.[3]

Each year 100,000 workers in the United States die from occupational diseases while nearly 400,000 new cases are reported.[4] Approximately nine million persons

---

Adapted with permission from Bunyan Bryant and Paul Mohai, *Race and the Incidence of Environmental Hazards: A Time for Discourse* (Boulder, Colo.: Westview Press, 1992).

each year suffer from severe work-related injuries.[5] African Americans, however, have a 37 percent greater chance of suffering an occupational injury or illness and a 20 percent greater chance of dying from an occupational disease or injury than do white workers. African-American workers are also twice as likely to be permanently or partially disabled due to a job-related injury or illness.[6]

African American workers in the United States constitute a major segment of the work force—12 percent of the total number of employed individuals in the United States—and represent 30 percent of all unionized workers. They constitute the highest proportional percentage of unionized workers in this country.[7]

Since the establishment of the Occupational Safety and Health Act of 1970, improvements in the general safety and health of workers have been made, particularly in the identification of hazards and the establishment of controls for toxic and cancer causing agents. African Americans and other minority workers, however, have not benefited from these improvements to the degree that white workers have. Among this highly unionized group within the work force, historically, the injury, disease, and death rates have been and remain disproportionately high as compared to their white counterparts in certain industries. Why have African American workers fared so poorly in the area of job-health safety?

Occupational injury and death statistics strongly indicate an unofficial policy of benign neglect for African American workers. They have not been the focus of federal agency policies established for "special targeted industries, occupations, substance, or exemptions for small business including agriculture."[8] Not surprisingly, occupational injury, illness, and death rates are significantly higher among African American workers as compared to white workers in certain dangerous industries. African American workers are in "double jeopardy" of loss of life and susceptibility to disease and injury because of racial discrimination. Because African Americans have been assigned a lower social status within American society, they have historically been relegated to the most hazardous jobs in dangerous industries with no possibility of advancement or improvement. Job discrimination, a pervasive fact of life for most African Americans, has ominous consequences for the health of African American workers. Available data, although limited, very clearly suggest that the concentration of African American workers in certain hazardous jobs within industries is responsible for excess disability and death rates among those workers.

This article reviews the somewhat limited data and investigates the effects of hazards in the work place on the health (i.e., the general health status) of African Americans,[9] specifically the extent to which the life expectancy of African Americans can be explained by occupational or job categories as opposed to intrinsic (i.e., diet, smoking, or drinking) racial or cultural differences between African Americans and whites.

Many theories are offered to explain diseases. Among them, social causation theories assert the idea that "we cannot understand disease incidence without looking at the social context in which people live and die."[10] This useful theoretical approach explains possible relationships among disease-causing agents as well as contributing factors within those relationships. From this premise, a study by Ellen Hall of health in inner cities identifies disease among low-income, inner-city minorities as physically, socially, and genetically induced.[11] All three causes provide insight into reasons why low-income, inner-city minorities (and other inner-city residents) contract disease at a much greater rate than more affluent urban, suburban, and rural residents. Hall's model is easily adapted to the questions of this research effort and works well in explaining why African American workers are at greater risk of injury, disease, and death than are white workers.

Socially induced diseases are those that result from social rather than physical, genetic, or environmental causes. Most important, these illnesses are often beyond the individual's control. For example, one of the ramifications of race may be sources of disease. In the case of African American workers, consequences of race could include job discrimination and job placement as well as numerous other stressors.

Physically induced diseases (including genetics) are those that result from physical rather than environmental or social causes. Physically induced diseases are those that occur because of intrinsic, racial, or cultural factors such as unhealthy diets and habits. Genetic factors are also classified as physical inducements of disease. Hall, however, does not include genetic inducement with physical inducement of disease. Stereotypes and myths about minority groups, as well as genetic differences between races, have often been used as sources of disease and as justification for singling out some groups for differential treatment. For example, in the case of African American workers, the myth concerning their ability to withstand hotter temperatures has been used as justification for their being assigned to the extremely dangerous coke ovens in the steel industry. Certain cultural habits, such as diets and smoking, are also used as explanations of disease and injury in African American workers.

Environmentally induced disease, for the purposes of this research effort, are separated from physically induced disease and are discussed separately. Environmentally induced diseases are those that result from environmental rather than physical or social causes and are caused by the quality of the environment in which we live. These include our homes, communities, and work environments. Most often, environmental exposure is involuntary, and individuals lack specific information on the hazards associated with their environments. For example, African American shipyard workers in coastal Georgia were found to have a disproportionately high lung-cancer death rate. This incidence of cancer was shown to be related to job exposure.[12]

HYPERTENSION

Stress and stress-related illnesses are identified as socially induced diseases. They can occur "from physical and psychological responses to a variety of social as well as pathological factors over which people have varying degrees of control."[13] Stress and stress-related illnesses represent a major health problem among African Americans since these diseases are often due to social factors beyond the individual's control.[14] Hypertension (elevated blood pressure) is a stress-related disease primarily understood as socially induced. An examination of hypertension among African Americans and a review of stress-related diseases within certain industries where significant numbers of African American workers are employed further illustrate the vulnerable position of African Americans in the workplace.

Hypertension is a major cause of organ damage and death in humans. It is believed to be the "body's response to its need to accommodate faster breathing and heart beat in response to stress and other factors."[15] Hypertension rates for African Americans of every age are nearly twice as high as those for whites.[16] Approximately six million of the twenty-five to thirty million Americans who suffer from hypertension are African American. Hypertension death rates are higher for both African American males and females than for white males or females.[17] Heart disease, which is hypertensive related, accounts for more deaths per year than any other single category. Although these statistics suggest an inherent or genetic propensity to hypertension among African Americans, hypertension rates for blacks who are not American and who reside in other countries are significantly lower than whites in those same countries.[18]

Explanations for this increased risk of hypertension and related diseases among African Americans range from cultural habits (e.g., eating, drinking, and smoking) to psychological stressors due to discrimination or genetics. A social factor contributing to this increased risk that is generally overlooked is stress in the workplace. *To what extent is stress or stress-related disease due to occupational environments?*

Available data lend support to the contention that occupational environments play a far more important role than is presently realized in causing stress-related diseases. Studies show that working conditions within certain industries contribute to the increased incidence of stress-related mortality and morbidity. African American workers are also more likely to be employed in stress-related job categories within these industries than are white workers. For example, studies of both iron and steel foundry workers and laundry and dry-cleaning industry workers show an increase in the incidence of stress-related mortality and morbidity among blacks as compared to white workers.[19]

These data seem to suggest that the social practice of discriminatory job placement has resulted in the assignment of African Americans to extremely hazardous jobs that are also stress inducing. African American workers, as compared to white workers, disproportionately suffer from stress-related illnesses and death. Although conclusive proof does not exist, these data suggest that discriminatory job placement based on race has harmed the health of African American workers in certain industries.

BLAMING THE VICTIM

Physically induced diseases—those that occur because of intrinsic factors such as diet, smoking or genetics—are often blamed on an individual's behavior and personal characteristics as the primary cause of disease. In occupational health and safety, some advance "victim-blaming" arguments to explain the higher injury and death rates of nonwhite workers. The basic argument, however, is that "modern industrial working conditions are so safe that if a worker gets hurt or sick it must be his or her fault and not the fault of the industry."[20] Consequently, the lower life expectancy and higher incidence of cancer and death rates incurred by nonwhite (especially African American) workers is often attributed to either bad habits or genetics. For example, Paul Kotin resurrected the "hyper-susceptibility worker" notion,[21] which shifts the focus from "what chemicals cause cancer" to "what people get cancer."[22] Kotin's argument begins with the reasonable assumption that "all biological organisms, including humans, vary in their response to external stimuli such as toxic substances or carcinogens." He then asserts that management has a "right" to select sturdier workers for riskier jobs. This viewpoint supports management's "right" to continue discriminatory job placement practices in some industries. For example, in the iron and steel industry, one study showed that 91 percent of all those working the coke ovens were African American and were exposed to (among other things) extreme heat.[23] The industry justifies this practice on the basis of the myth that, "African Americans absorb heat better."[24] Similar practices based on myths were found within the electronics industry, where "dark-skinned" minority workers were regularly assigned to jobs using caustic chemicals because skin irritations resulting from job exposure are not as pronounced on dark skin as they are on white skin. Consequently, "dark-skinned" workers should have had fewer complaints than their white counterparts whose skin irritations were more noticeable.[25]

A second blame-the-victim tactic shifts the responsibility for occupational injury and death from uncontrolled exposure in the workplace to the worker's own insidious lifestyle.[26] For example, management often assigns the cause for the high incidence of lung cancer found among workers in some industries to smoking habits rather than to overexposure to dusts and chemicals in the workplace. Smoking definitely increases the susceptibility of workers exposed to dust and chemicals (e.g., asbestos, rubber, and steel workers), but the risk to nonsmoking workers in such industries is also greater than that of persons who do not work in such environments.[27]

African American males have the highest reported incidence of lung cancer. The death rate for lung cancer among African Americans is about 20 times higher than it was in the early 1950s. Although smoking certainly accounts for some of the increase in lung cancer among African American males, this alarming increase cannot be attributed to smoking alone. The excess risks are more likely due to environmental factors, including occupational exposures. Only recently are we uncovering occupationally induced cancer problems that have existed for years. Present data reveal that a significant number of African American workers in specific industries have been assigned to the most hazardous jobs that also exposed them to numerous now-known carcinogens.[28] The degree to which excess risks of lung cancer in African American males may be attributed to occupational exposure is reflected, for example, in the fact that the highest incidence of lung cancer among African American males is in Pittsburgh, Pennsylvania. This is not surprising, given that a significant number of African American males in Pittsburgh are employed in the most hazardous jobs in the steel industry, which exposes them to known carcinogens.

Blaming the victims is an industry management tactic used to justify inaction. Myths or racist stereotypes camouflage discriminatory job placement practices resulting in the purposeful exposure of African American workers to hazardous work conditions. Equally insidious is the victim-blaming tactic employed by industry that blames occupational illness and death on intrinsic, racial, or cultural characteristics, including alleged genetic deficiencies. This rationale then absolves the industry of any blame for disease or injury rates among workers.

OCCUPATIONAL CANCER

Environmentally induced diseases are those that occur due to exposures in the environment. The work environments of some industries have already been shown to be extremely hazardous for workers, especially African American workers. Not surprisingly, cancer rates for African Americans are increasing in epidemic proportion, and African American workers experience disproportionately high cancer incidence and death rates. Although in 1949 the reported cancer rate for African Americans was 20 percent lower than the rate reported for whites, by 1967 the number of deaths from all cancer increased twice as rapidly among African Americans as it did among whites. Presently, African American males have the highest-reported incidence rate for all cancers combined. African Americans are experiencing a growing cancer epidemic with a 25 percent increase in the cancer rate since 1980. The American Cancer Society data show that the cancer mortality rates for blacks and whites were practically the same since the early 1960s.[29] However, cancer death rates in whites have subsequently increased by 10 percent, while the rate for African Americans has increased by an astounding 40 percent. Many factors have been cited as possible contributors to the high incidence of cancer among African Americans, including smoking, diet, and genetics. The relationship between occupational exposures and cancer has received only minimal attention by the scientific community. However, available data tend to support the relationship between some work environments and the incidence of cancer among workers.

The exposure of African American workers to many of the chemicals that are now known as carcinogens began in the early 1900s, when large numbers of Southern African Americans migrated to urban industrial areas to work. The status and nature of the jobs, however, were no different from those relegated to them in the South. African American workers were hired for the worst jobs, most often the most strenuous and hazardous, that also exposed them to chemicals that are now known to cause cancer.[30] Studies of industries where large numbers of African Americans have worked in jobs using or producing carcinogens suggest that this exposure is responsible for some excess risk of cancer among African American workers.[31]

Illustrations of industry exposure of African American workers to possible cancer-causing agents include the following:

1. The National Cancer Institute conducted a study of laundry and dry-cleaning workers and found that African Americans had higher death rates from cancer of the liver, lung, cervix, uterus, and skin.[32]

2. A 1978 cancer mortality study of coastal Georgia residents found African American shipyard workers to have a lung cancer death rate two times higher than expected.[33]

3. A 1946–1950 U.S. Public Health study of chromate workers found that the respiratory-cancer mortality rate for all workers was twenty-nine times higher than expected. However, the actual-to-expected respiratory-cancer death ratio was 14.29 for whites and 80.00 for blacks.

4. A nine-year study of 59,000 steelworkers (62 percent of all U.S. males working in basic steel production) revealed that nonwhite workers (mostly African American) in the coke plant experienced double the expected death rate from malignant neoplasms, due mostly to cancer of the respiratory system.

These data suggest that the excess risk of cancer that exists for African American workers, as compared to white workers, may be due to greater exposure of African American workers to carcinogens in the work place. Moreover, the placement of more African Americans than whites in the most dangerous jobs in certain industries may account for this overexposure.

CONCLUSION

Numerous factors contribute to the health status of African Americans. Occupational exposures, however, receive little attention. Statistics on occupational safety and health are generally lacking, and race-specific data is even harder to find. However, African American workers represent over 15 percent of the total work force in nearly thirty-three occupational categories. Unfortunately, a large percentage of the African American work force remains overrepresented in low-pay, low-skill, high-risk blue-collar and service occupations. Moreover, African Americans are concentrated in certain industries, many of which have above-average injury and illness rates, including laundry and dry-cleaning, tobacco manufacture, fabric mills, smelters, hospitals (as orderlies and attendants), and service industries. Available data on job placement patterns within certain industries suggest an even more serious health threat for African American workers. African American workers are also overrepresented in the dirtiest and most hazardous jobs in certain industries. Generally, African American workers are relegated by discriminatory employment practices to the least-desirable jobs.

Historically, racist attitudes or practices have exacerbated health and safety problems. For example, the Gauley Bridge disaster (1930–31) in West Virginia was responsible for the disability of fifteen hundred workers and the death of five hundred mostly African American workers, who were recruited to tunnel through a mountain with a high silica content. Overexposure to this substance usually causes a chronic lung disease. In the Gauley Bridge incident, 169 African American men literally dropped dead and were hurriedly buried on the spot. The workers, who earned about thirty cents an hour, were not told of the known hazards or given protective breathing devices. The disaster was not uncovered until 1935 and resulted in the amendment of the West Virginia compensation law, but not soon enough to benefit the dead, the disabled, or the family members.[34]

Similarly, in 1969 the textile industry denied evidence that exposure to cotton dust could cause byssinosis ("brown lung" disease). The general sentiment of the industry on this matter was reflected in an editorial in the industry's journal: "We are particularly intrigued by the term 'byssinosis,' a thing thought up by the venal doctors who attended the last International Labor Association meeting in Africa where inferior races are bound to be inflicted by diseases more superior races defeated years ago."[35]

Today, African American workers are still relegated to the dirtiest and most hazardous jobs within certain industries. Evidence supports the contention that working conditions in certain industries employing significant numbers of African Americans are contributing to the increased incidence of injury, disease, and death among African Americans as compared to white workers. Moreover, there is a great disparity between the general health status (as represented by important health indicators such as life expectancy, incidence of disease, and death rates) of African Americans as compared to whites. Available data, although limited, suggest that occupational factors play a significant role in the causation of major disease and health problems among African Americans. That African American workers are at special risk of loss of life and susceptibility to disease and injury because of discriminatory job placement practices is strongly supported by these data. *A job should not be a death sentence.* However, African American workers, due to their race and resultant discrimination, are relegated to jobs with increased risk of injury, disease, and death. Race and occupation combine to place African Americans in double jeopardy of loss of life and susceptibility to disease. The effects on the general health status of African Americans seems ominous and is even more distressing in light of the inaction on our government's part to address this problem. Even more disturbing, much evidence supports the notion that African Americans are generally not aware of the dangers in their work environment.

In sum, arguments that place the blame for the excess risk of injury, disease, and death of African American workers on physical factors (such as their "hypersusceptibility") or those that place the blame for the excess risk of injury, disease, and death of African Americans in general on intrinsic factors (such as diets, smoking, and genetics) are generally disputed by the research. These data suggest that it is unlikely that physical or genetic characteristics are solely the blame for the increased health risk among African Americans. It seems much more likely that environmental causes, including the work environment, contribute far more to this increased risk of injury, disease, and death among African Americans than has been realized.

NOTES

1. U.S. Department of Health and Human Services, *Report of the Secretary's Task Force on Black and Minority Health* (Washington, D.C.: Government Printing Office, 1985), 2.
2. Ibid., 51.

3. See Nicholas A. Ashford, "Crisis in the Workplace: Occupational Disease and Injury," *Report to the Ford Foundation* (Cambridge, Mass.: MIT Press, 1976); Morris E. Davis, "Occupational Hazards and Black Workers," *Urban Health* (August 1977): 16-18; and "The Impact of Workplace Health and Safety on Black Workers: Assessment and Prognosis," *Labor Studies Journal* 4 (Spring 1981): 29-40; Samuel S. Epstein, Lester O. Brown, and Carl Pope, *Hazardous Waste in America* (San Francisco: Sierra Club Books, 1983); and Robert Bullard and Beverly Hendrix Wright, "The Politics of Pollution: Implications for the Black Community," *Phylon* 47 (1986): 71-78.

4. U.S. Council on Environmental Quality, *Environmental Quality: The Tenth Annual Report* (Washington, D.C.: Government Printing Office, 1980); and Ray Elling, *The Struggle for Workers' Health: A Study of Six Industrialized Countries* (New York: Baywood Publishing Co., 1986).

5. Congressional Quarterly, Inc., *Environment and Health* (Washington, D.C.: Congressional Quarterly, 1981).

6. Urban Environment Conference, *Taking Back Our Health* (Washington, D.C.: Urban Environment Conference, Inc., 1984).

7. Davis, "Impact of Workplace Health," 1981, 29.

8. Ibid., 30.

9. Ibid.

10. Ellen Hall, *Inner City Health in America* (Washington, D.C.: Urban Environment Foundation, 1979), 13.

11. Ibid.

12. William Blot et. al., "Lung Cancer after Employment in Shipyards During World War II," *New England Journal of Medicine* 21 (December 1978).

13. Hall, *Inner City Health*, 29.

14. Joseph Eyers, "Hypertension as a Disease of Modern Society," *International Journal of Health Services* 5 (1975): 547.

15. Hall, *Inner City Health*, 32.

16. Jay Weiss, *Psychological Factors in Stress and Disease* (San Francisco: W. H. Freeman, 1976), 165.

17. Hall, *Inner City Health*, 32.

18. Ibid.

19. H. Rockett and C. Redmond, "Long-Term Mortality Study of Steelworkers and Mortality Patterns Among Masons," *Journal of Occupational Medicine* 18 (August 1976): 541–545; Aaron Blaire, "Causes of Death Among Laundry and Dry-Cleaning Workers," *American Journal of Public Health* 69 (May 1979): 509; and Davis, "Impact of Workplace Health," 33.

20. Samuel S. Epstein, *The Politics of Cancer* (San Francisco: Sierra Club Books, 1978), 395.

21. Paul Kotin, an address to the American Occupational Medicine Association (delivered in Denver, Colo., 1977).

22. Epstein, *Politics of Cancer*, 395.

23. Lloyd Williams, "Long-Term Mortality of Steelworkers: Respiratory Cancer in Coke Plant Workers V," *Journal of Occupational Medicine* 13 (1971): 55.

24. Davis, "Occupational Hazards," 17.

25. Ibid.

26. Epstein, *Politics of Cancer*, 396; Davis, 33; Urban Environment Conference, 5.

27. Epstein, *Politics of Cancer,* 366; Urban Environment Conference, 5.
28. Davis, "Occupational Hazards," 17; Davis, 33; Epstein, *Politics of Cancer,* 396.
29. American Cancer Society, *Cancer Facts and Figures: 1986* (New York: American Cancer Society, 1986).
30. Davis, 33-36.
31. Urban Environment Conference, 5-6; Davis, 34; Hall, *Inner City Health,* 19-28.
32. Blaire, "Causes of Death," 509.
33. See Blot et. al., 1978.
34. Davis, "Occupational Hazards," 17.
35. American Textile Reporter, 1969; Davis, "Occupational Hazards," 17.

16

Farm Workers at Risk

Cesar Chavez

WHAT IS THE worth of a man or a woman? What is the worth of a farm worker? How do you measure the value of a life? Ask the parents of Miriam Robles.

Miriam Robles was a ten-year-old girl when she died in 1992 of leukemia, the third farmworker child to succumb to leukemia in the small Tulare County town of Earlimart, in California's Central Valley. The disease had already claimed the lives of little Jimmy Caudillo and Monica Tovar. Dozens of children have been stricken by cancer and birth defects in Earlimart and other nearby farmworker communities.

Miriam's parents and the parents of Jimmy Caudillo and Monica Tovar have two things in common. They are all farm workers who pick grapes. And they are all constantly exposed to pesticides, from the vineyards in which they work, from the heavily and repeatedly sprayed fields surrounding their homes, and from pesticides that pollute irrigation water and groundwater.

Pesticides, to be sure, are not the only form of oppression plaguing farm workers. Farm labor contractors, middlemen who supply labor to growers, force workers to live in crowded, dirty labor camps, where they are often stacked in three-high bunks. Utility costs, even for those who must live and cook outdoors, get deducted from their paychecks. Many grape workers have to pay contractors a fee for a ride to work, even if they use their own cars or live in camps near the vineyards.

In the vineyards, workers are frequently compelled to labor through legally guaranteed lunch and work breaks to meet job quotas. Daily tallies are kept. Legally required portable toilets are often far from where people work. Crew leaders time those who use them. Toilets are moved over rough terrain so that water with human waste dirties the walls and floors, compelling farm workers to relieve themselves in the vineyards.

Cesar Chavez died on April 22, 1993.

Instead of earning minimum wage, farm workers are often paid "by the piece," so many cents for each box of grapes they pick. As a result, workers often make as little as $2 an hour, even though such a rate violates federal and state labor laws.

Farm workers often say that growers and labor contractors treat them as if they are animals. These miserable conditions are the rule and not the exception in vineyards, orchards, and fields throughout California and across America. They have remained essentially unchanged for generations.

But even more important to farm workers than the better wages and freedom from exploitation that United Farm Workers (UFW) contracts bring is protecting workers—and consumers—from systematic poisoning through the reckless use of agricultural toxics.

There is nothing we care more about than the lives and safety of our families. There is nothing we share more deeply in common with the people of North America than the safety of the food we all rely upon.

The chemical companies that manufacture the pesticides and the growers who use them want us to believe that they are the health-givers, that because of pesticides people are not dying of malaria and starvation. They have convinced the politicians and the government regulators that pesticides are the cure-all, the key to an abundance of food. So they don't ban the worst of these poisons because some farm workers give birth to children who contract cancer or to babies who are born with deformities. They don't imperil millions of dollars in profits today because, some day, some consumers might get cancer. They allow all of us who place our faith in the safety of the nation's food supply to consume grapes and other produce that contain residues from pesticides that cause cancer and birth defects. We accept decades of environmental damage these poisons have brought upon the land. The growers and the chemical companies, the politicians and the bureaucrats, all say that these are acceptable levels of exposure. Acceptable to whom?

Acceptable to Miriam Robles' parents and the parents of Jimmy Caudillo and Monica Tovar? Acceptable to Felipe Franco, a young boy in Delano, California, who was born without arms or legs after his mother worked in the vineyards throughout her pregnancy?

Acceptable to all the other farm workers and their children who have known tragedy from pesticides? Acceptable to the 300,000 farm workers who are poisoned each year in the United States, according to a 1985 study by the World Resources Institute?[1] Or the eight hundred to one thousand who die each year from exposure to pesticides, according to a study by the U.S. Food and Drug Administration?[2]

There is no acceptable level of exposure to any chemical that causes cancer. We cannot tolerate any toxic substance that causes miscarriages, stillbirths, and deformed infants.

Isn't that the standard of protection you would ask for your family and your children? Isn't that the standard of protection you would demand for yourself? Then why not for farm workers?

Compared with other jobs, farm workers are one of the least-protected groups in the nation. They are specifically excluded, either totally or partially, from health and safety standards under the federal Occupational Safety and Health Act as well as the Fair Labor Standards Act. Many are excluded from state worker compensation and unemployment insurance laws. Thus we lack effective legal remedies. Growers are not even required to tell workers the specific chemicals being used or to provide protective clothing. For one of the most hazardous occupations in the nation, three and one half million farm workers live under a double standard.

Pesticide poisoning of our children is perhaps the ultimate form of oppression. But, as already noted, farm workers and their families today endure many burdens, products of the greed and racism that overtook California and the nation during the 1980s and '90s—although the law of the jungle has prevailed for at least one hundred years in California fields. Dozens of labor groups tried and failed to organize farm workers, from the International Workers of the World (the "Wobblies") at the beginning of the twentieth century to the old American Federation of Labor in the 1940s and '50s. New Deal reforms passed in the 1930s excluded farm workers but conferred the right to organize on industrial workers.

Rural judges broke strikes with sweeping injunctions. Police agencies and armed vigilantes crushed walkouts by breaking strikers' heads. Strikers were beaten. Farm workers were killed.

In 1965, our infant farm workers' union led a major strike against Delano-area grape producers. That walkout was also headed for the same fate as its predecessors, until we tried something different: We asked consumers across North America to boycott California table grapes.

In 1992, the United Farm Workers marked its 30th anniversary. It has been almost twenty-seven years since our union cause first touched the hearts and consciences of people across America and around the world by letting them know about the abuses suffered by farm workers and their families.

In 1967, the farm workers dramatically transformed the simple act of refusing to buy fresh table grapes into a powerful statement against unfairness and injustice. The grape boycott was a hallmark of the 1960s and '70s. It rallied millions of Americans to the cause of migrant farm workers. And it worked. The first real collective-bargaining relationship was established between an agricultural employer and farm workers when we signed our first union contract with a Delano-area grape grower.

A law enacted in California in 1975 guaranteed farm workers the same rights industrial workers had won four decades before. That law, pushed through by then-Governor Jerry Brown, protected the right of farm workers to organize and to decide, in secret-ballot elections, whether they wished to be represented by a union. Since 1975, the United Farm Workers has won 423 union elections. Despite long delays and bureaucratic inertia by the California Agricultural Labor Relations Board during the Brown administration, farm workers began to make progress, signing contracts with growers.

By the early 1980s, tens of thousands of farm workers enjoyed the benefits of UFW contracts: the first comprehensive medical coverage for farm workers and their families under the self-insured Robert F. Kennedy Farm Workers Medical Plan; benefits paid under the nation's first and only pension plan for retired farm workers; paid holidays and vacations; mandatory rest periods; clean drinking water, hand-washing facilities, and rest rooms in the fields; prohibitions on job discrimination against female farm workers and an end to sexual harassment by foremen and supervisors; abolition of the farm labor contractors who directly cheat and exploit farm workers; and the first cost-of-living guarantee in a union contract for farm workers in U.S. history.

Although UFW contracts in 1967 provided protection from dangerous pesticides, most farm workers still remained unprotected in 1982, when corporate growers gave the gubernatorial election campaign of Republican George Deukmejian more than $1 million.

All but the most biased observers now admit that under Deukmejian and his successor, Republican Governor Pete Wilson, the farm-labor law became a cruel hoax. Corporate agribusiness gave Deukmejian and Wilson millions of dollars in campaign contributions. They paid back their debt to the growers with the blood and sweat of California farm workers.

Charges filed by farm workers against growers have been dismissed without being investigated. Cases where the state Agricultural Labor Relations Board and the courts ordered growers to pay farm workers millions of dollars in damages for breaking the law were settled for as little as ten cents on the dollar—in violation of state and federal labor-law rules. State civil servants who insisted on enforcing the law were intimidated by political appointees of the governor.

Rene Lopez, a nineteen-year-old farm worker and UFW member at Sikkema Farms dairy, was shot to death by grower agents shortly after voting in a union-representation election near Fresno, California, in 1983.

Before the election, the state farm labor board refused to act on complaints from farm workers that grower agents were threatening workers with guns. After Rene was killed, security guards at the company began carrying billy clubs, rifles, pistols, and shotguns. The governor's political appointees refused to seek a court order barring these weapons. The UFW had to go to court to get the injunction.

Thousands of farm workers who were beaten, threatened, fired, and blacklisted for organizing and supporting the UFW still don't have what they sacrificed for: a union contract. A 1991 state report showed that growers owed money to 11,274 farm workers who lost income due to antiunion discrimination. Only 348 of them were working under union contract.

Critics of our union cause, especially growers, say that the United Farm Workers doesn't do enough to organize farm workers. The issue isn't organizing; it's negotiating contracts in good faith, something most growers have stopped doing. The UFW still represents about 85,000 farm workers who voted for the union but can't get contracts.

Many growers use sophisticated corporate reshuffling schemes to avoid their legal obligation to bargain after workers vote for the UFW. One day, Company A announces that it has gone out of business. The next day it reemerges as Companies B, C, and D. The same managers farm the same land with the same equipment, the same supervisors, and foremen. The only difference is a whole new nonunion work force. The state farm labor board does nothing.

The progress that some farm workers achieved by the early 1980s only highlights the miserable conditions that have returned to the fields: child labor, pesticide poisonings, sexual harassment of female workers, forcing workers to labor faster, vicious discrimination against older farm workers, and rampant violation of state and federal minimum wage and hour laws.

Farm workers must live out in the open in caves, canyons, and under trees. Workers sleep outside in overcrowded farm labor camps, while labor contracts deduct utility costs from their paychecks. These outrages once again led the UFW to our court of last resort: We asked the American people to support a new boycott of California table grapes.

The growers blame us for resorting to the boycott instead of organizing farm workers. But agribusiness makes it impossible for farm workers to organize and win contracts under the law. The UFW may be the only union in America criticized by employers for not doing enough to organize their workers.

Many of these same growers also spent millions of dollars to help kill Proposition 128, California's 1990 Big Green initiative supported by environmental groups and the UFW, which would have protected California's last stands of privately held redwoods and banned cancer-causing pesticides.

Growers and other opponents of Proposition 128 would be mistaken to view the election results as a rejection of the need for reforms, such as the protection of old-growth forests and restrictions on the use of pesticides that produce birth defects.

The election returns also included an important lesson for those interested in reforming our state and nation: Electoral politics is not the only medium of change. More than ever, direct confrontation and grassroots appeals—principal tactics of the environmental, antiwar and other popular movements in the 1960s and '70—are still viable alternatives in the 1990s.

Americans truly interested in working for social change cannot expect much from the political process for redress of their grievances and for solutions to their problems, especially if the people seeking redress happen to be farm workers, minorities or the poor.

This loss of faith in the political process should not be confused with the impatience that the young sometimes express when they realize the world will not be changed as soon as they would like. For forty years, even before the farm workers' union was founded, we have helped people become citizens by registering people to vote and organizing people to go to the polls and cast their ballots.

All unions must get involved in politics so that the gains working people win at the bargaining table are not lost in the legislatures. The UFW was no exception. The farm workers, perhaps more than most union members, supported candidates, walked door-to-door, turned out the vote, and made contributions from the money they earned laboring in the fields.

Yet the process of political involvement has become so expensive and the results so minuscule that we question our faith in the system. We have not failed the process; the process has failed us. Our commitment has not been respected. Our loyalty has not been returned.

The prospects of getting results for the poor are so meager, and the process of citizenship and voter registration made so difficult and intimidating, especially for non-English–speaking minorities, that masses of people fail to participate.

That is why, in a state as ethnically and racially diverse as California, the electorate is so much older, whiter, wealthier, and more conservative than the population as a whole, even the eligible voting population.

Solutions are not likely to be found through public policy, which requires that you place your fate in the hands of the politicians. Solutions can be achieved through public action, taking matters into your own hands by taking your case directly to the people.

When you transfer your allegiance from public policy to public action, the polls never close, your supporters can vote more than once, and you don't need a majority to win.

Corporate special interests can marshal tens of millions of dollars to kill citizen-sponsored initiatives such as the Big Green initiative. But the challenge from a simple boycott can render their money and clout impotent.

Our adversaries often mistakenly apply the rules of electoral politics to boycotts. Not long ago, the head of the California Table Grape Commission announced that his group was halting a multimillion dollar advertising campaign to counter our current grape boycott because polls showed a majority of consumers weren't supporting the boycott.

But boycotts don't need 50 percent plus one to win. A boycott that gets 5 percent of consumers to cooperate is doing very well. A 10 percent drop in sales can be devastating. And those consumers can vote for your cause every time they go shopping.

Most high-volume supermarket chains operate on profit margins of 1 or 2 percent. When grape-boycott volunteers turn away thousands of customers, as they have done in front of Southern California Vons stores, it costs supermarkets millions of dollars.

Our opponents commonly make another key miscalculation about boycotts. They argue that it's been nine years since the grape boycott began again, and it still hasn't been won.

When we win isn't important. The rich have money and the poor have time. We don't have to win this year or next year or even the year after that.

Until the boycott takes its toll, we will not give up. We have nothing else to do with our lives except to continue in this nonviolent fight.

In political campaigns, you race against time to get your message out, and you are always dramatically outspent. With boycotts and other forms of public action, time becomes your ally. In the end, it can be a more powerful force than all the money that the corporations can muster.

All my life, I have been driven by one dream, one goal, one vision: To overthrow a farm-labor system in this nation that treats farm workers as if we are not important human beings. Farm workers are not agricultural implements or beasts of burden to be used and discarded.

That dream, born in my youth, was nurtured in my early days of organizing. It has flourished; it has also been attacked. My motivation comes from my personal life, from watching what my parents went through as migrant farm workers in California in the 1930s and '40s.

That dream grew from my own experience with racism, with hope, with the desire to be treated fairly and to see my people treated as human beings, not as chattel.

It grew from anger and rage, emotions I felt forty and fifty years ago when people of my color were denied the right to see a movie or eat at a restaurant in many parts of California. It grew from the frustration and humiliation I felt as a boy who couldn't understand how the growers could abuse and exploit farm workers when there were so many of us and so few of them.

Later, in the 1950s, I began to realize what other minority people had discovered: The only answer, the only hope, was in organizing. All Latinos, urban and rural, young and old, are connected to the farm workers' experience. We all lived through the fields, or our parents did. We shared that common humiliation.

How could we progress as a people, even if we lived in the cities, while the farm workers—men and women of our color—were condemned to a life without pride? How could we progress as a people while the farm workers, who symbolized our history in this land, were denied self-respect? How could our people believe that their children could become lawyers, doctors, judges, and professional people while permitting this shame and injustice to continue?

Those who attack our union often say, "It's not really a union. It's something else—a social movement, a civil-rights movement; it's something dangerous." They're half right. The UFW is first and foremost a union. But we have always been something more than a union, although we've never been dangerous if you believe in the Bill of Rights.

We attacked that historical source of shame and infamy that our people in this country lived with not by complaining or by seeking handouts. We organized! Farm workers acknowledged we had allowed ourselves to become victims in a democratic society where majority rule and collective bargaining are supposed to be more than political rhetoric. And, by addressing this historical problem, we created confidence, pride, and hope in an entire people's ability to create the future.

The UFW's survival was not in doubt after the union became visible, when Latinos started entering college in greater numbers, began running for public office in greater numbers, and started asserting their rights on a broad range of issues in many communities.

The UFW's existence signaled to all Latinos that we were fighting for our dignity, challenging and overcoming injustice, and empowering the least educated and poorest among us. The message was clear: If it could happen in the fields, it could happen anywhere—in the cities, in the courts, in the city councils, and in the state legislatures.

I have met and spoken with thousands of Latinos from every walk of life and from every social and economic class. What I hear most often from Latinos, regardless of age or position, and from many non-Latinos as well, is that the farm workers gave them the hope that they could succeed and the inspiration to work for change.

Latinos across California and the nation who don't work in agriculture are better off today because of what the farm workers taught people about organization, pride and strength, and control over their own lives.

Tens of thousands of the children and grandchildren of farm workers, and the children and grandchildren of poor Latinos, are moving out of the fields and the barrios. That movement cannot be overturned. Once social change begins, it cannot be reversed. You cannot uneducate the person who has learned to read. You cannot humiliate the person who feels pride. You cannot oppress the people who are not afraid anymore.

Boycott grapes!

NOTES

1. Robert F. Wasserstrom and Richard Wiles, *Field Duty: U.S. Farmworkers and Pesticide Safety* (Washington, D.C.: World Resources Institute, 1988): 3.
2. See 52, no. 84, Federal Regulations, 16.065 (May 1, 1987).

17

Work: The Most Dangerous Environment

Charles Noble

FOR MANY PEOPLE, work is the most dangerous environment. Workplace hazards are among the highest causes of preventable injury and death in the United States.[1] Yet, in a society obsessed with disease and dying, remarkably little is done to anticipate and avert these unnecessary tragedies, even though laws have been enacted, notably the Occupational Safety and Health Act of 1970, and agencies created, including the Occupational Safety and Health Administration (OSHA).

Opponents of workplace regulation contend that work is inherently dangerous: Beyond a certain point—a point that they think we have already reached—government's effort to protect workers is hopelessly utopian. Little more can be accomplished, they argue. We will only hurt productivity and slow economic growth.

But neither argument is true. While the workplace is always potentially dangerous, elevated levels of life-threatening risks are not inevitable. Attention to the impact of new technologies, to the design of the production process, and to worker training — all make it less likely that employees will suffer accidents and injuries. Nor must a safe economy be a sick economy. Some of our most effective competitors, such as Germany, take worker rights much more seriously. The real constraints on occupational safety and health reform lie elsewhere, in the broader social, economic, and political relationships that shape what can be and is done to reform the workplace in the United States.

THE FIVE PREREQUISITES OF EFFECTIVE REFORM

Five conditions must be met if any society is to deal intelligently and responsibly with the hazards of work.

Good Science

There is an irreducible scientific component to all environmental reform: Effective responses depend on the development and dissemination of technical information—how to recognize risks, how to understand when and why they are likely to occur, and how to change the production process to avoid them. A wholesale assault on asbestos-related diseases and deaths, for example, waited until research conducted by Dr. Irving Selikoff conclusively demonstrated the harmful effects of this toxic substance. But there are thousands of potentially hazardous substances in the workplace. In the face of corporate resistance to regulation, it is difficult to conduct the research necessary to control each and every hazardous substance.

For scientists, career rewards lie elsewhere, working for companies interested in turning scientific research into profit-making applications, not in exposing dangers at work. And public and private agencies rarely fund the large-scale studies necessary to establish the origins of workplace diseases.

Unions are an alternative source of funding, but in the United States they do not have the resources needed to sponsor the necessary research. As a result, the labor movement has depended on the efforts of individual doctors and scientists committed to public health and worker safety. Many crusading researchers have stepped into this breach and made an enormous difference, beginning with Dr. Alice Hamilton, who conducted the first systematic inquiry into workplace diseases in the United States in the early twentieth century. But individual scientists have difficulty sustaining this kind of research effort. Family practitioners rarely have the training or perspective to see the link between what someone does for a living and the complaints they bring to the doctor's office. Individual research scientists lack the resources to fight the intense corporate campaign that usually follows any effort to publicize workplace hazards.

Sometimes professional associations have challenged corporate-dominated science. The American Conference of Governmental Industrial Hygienists has challenged many of the positions taken by corporate-sponsored professionals. The National Association for Public Health Policy maintains an Occupational Health Division that focuses attention on workplace safety issues. But, too often, professional associations are run by the same doctors and scientists who depend on corporate employment and research grants. As a result, these professional associations are usually reluctant to rock the boat by investigating workplace hazards or by publicizing what is already known about occupational safety and health. The best way to surmount the resource problem is to develop well-funded, public institutions with the organizational capacity to do scientific research about occupational accidents and illness, and to mobilize the political will to bring these matters to public attention.

Employer Compliance

Corporate resistance to workplace reform takes us to the second prerequisite of change: the commitment by those who run businesses in the United States, whether corporate executives or the owners of small businesses, to worker protection. Their cooperation is essential because, in market societies, they have the best information about what goes on at work, they have the day-to-day control over the labor process, and they are the ones who chose new technologies. Thus, it is absolutely essential that workplace reformers gain at least their acquiescence.

At a minimum, employers must acknowledge that workplace hazards are pervasive. More broadly, they must make the control of hazards one of the first principles of workplace organization. Where employers do this, workers are much more likely to be protected. Where employers do not, precious time and scarce money will be wasted fighting their resistance to change.

Of course, some companies already take occupational safety and health seriously; Johnson & Johnson is a well-known example of effective worker training and management attention to potential hazards. But too many firms downplay the problem and do not internalize this goal. For them, safety and health is another burdensome cost, to be avoided where possible. Or it is a "labor-management" problem, one that challenges managers to demonstrate their control over the workplace. Employer resistance is especially hard to overcome in a lax regulatory environment, where employers can often avoid the costs of cleaning up their work places.

In the face of this resistance, workplace reformers have three levers: pressure from below by workers, pressure from broader constituencies committed to worker rights to health and safety, and pressure from government.

Pressure from Below

Worker pressure is a remarkably efficient way to change employer practices. As long as the political system is democratic, and the option of wholesale repression of dissent is foreclosed, well-organized workers can force employers to deal with their concerns. Employers may resist at first, and continue to complain later, but eventually they will recognize workplace reform as a necessary cost of doing business. Worker pressure is also a remarkably efficient way to force government to respond. Political leaders generally prefer not to risk losing the votes of a large, well-organized labor movement. For these reasons, the most reformist states in Western Europe are those with the highest union density (i.e., the percentage of the labor force in unions). In Austria, Norway, Denmark, and Sweden—all leaders in occupational safety and health policy—at least 60 percent of the labor force is unionized.

In the United States, in contrast, unions are unusually weak. Less than one-fifth of U.S. workers are unionized. And far from calling the shots in Washington, D.C., the labor movement is fighting for its organizational life. As a result, neither the AFL-CIO nor many individual unions have the capacity to make occupational safety and health a priority issue—either on the shopfloor, at the bargaining table, or in the halls of Congress.

Workers receive support from other reform movements, including the public-interest, consumer, public-health, and environmental movements. Since the 1960s, many activists in these movements have seen the need to connect struggles over th "indoor" and "outdoor" environments. Many environmentalists now recognize ho polluted work sites damage the surrounding environment and the possibility forging coalitions around worker and community right to know about the use an control of toxic materials. As a result, coalitions between labor and these oth movements regularly form to lobby on issues of mutual concern. These movemen were particularly successful in the late 1970s and early 1980s in winnin right-to-know legislation in the state legislatures.

But today all movements are stretched thin as they work to defend existin environmental regulatory protections against successive conservative presidenti administrations. In some cases, environmentalists' elitist attitudes about workers, an fights between workers and environmentalists over the impact of environment regulation on jobs, have divided these constituencies and limited their politic strength. Conflicts between environmentalists and workers over the impact of loggin in old-growth forests in the Northwest on the fate of the spotted owl shows th divisiveness of these issues.

A Culture of Workers' Rights

The fourth prerequisite for effective workplace reform is a societal consensus on th rights of workers to safe and healthy workplaces. For workplace reform to become reality, the political culture must incorporate the idea that the costs of economi growth and change are *social* costs and that workers should not be asked to bear ther alone.

Unfortunately, while Americans endorse the idea of occupational safety and healt regulation, as they endorse almost all government regulation of business, they do nc appear to have taken the cause of worker safety fully to heart. Why isn't the publi more enthusiastic about this issue? Some people may assume that since workers ca choose to quit their jobs or protest if conditions are really bad, they must be *choosin* to work under dangerous conditions, perhaps for the better pay that these jobs offe Undoubtedly, some people think that workplace safety and health is a union issu since many Americans do not like unions, this tempers their enthusiasm for workplac reform. And other people, better disposed to unions, probably assume that the union will take care of their own.

All of these assumptions are incorrect. Some of the most hazardous jobs are also th worst-paid jobs, as farm workers know. Nor is occupational safety and health simpl a union issue. With their backs against the wall, the labor movement is simpl unprepared to fight this alone. Nonetheless, subjected to a constant barrage o corporate-inspired propaganda against unions and their special interests, too man Americans are prepared to let it try.

Coalition work among the various reform movements, in particular environmenta and labor activists, is essential if worker protection is to become an accepted part o the political culture. The idea of environmental protection appears to have alread achieved this privileged status. Like education, it has become what political scientists call a valence issue: Everyone is for it, and no one is against it, although people can argue bitterly over how to accomplish it. Environmentalists must now lend some of their credibility to the issue of worker safety and health, even while they struggle with union movements over other issues.

Responsive Government

The final precondition to effective workplace safety and health is a government willing and able to turn popular support for reform into a serious program. This means two things. First, government must be sufficiently responsive to the popular will to overcome employers' opposition to effective regulation. A supportive political culture, broad coalitions, and a strong labor movement are all part of the answer, but other political reforms, including but not limited to campaign finance reform, are also necessary. Second, government must learn to apply its regulatory power efficiently and effectively. Today we are far from the kind of industrial capitalism where mine and factory inspectors could solve these problems by fining recalcitrant employers (though even this rarely happened).

Government needs to develop regulatory institutions appropriate to a large, diverse, and complex postindustrial economy. There are still mines and factories to inspect. But increasingly the cutting edge of regulation empowers employees to take greater responsibility for their own well-being, and to develop proactive programs that can anticipate the harmful effects of new technologies. It also requires steering companies, industries, and the entire economy in more responsible directions—for example, away from pesticide-intensive agriculture to organic farming and away from nuclear power and fossil fuel to alternative energy sources.[2]

Worker empowerment does not mean asking employees to listen to speeches about how "safety pays." It involves giving them rights to take part in the decisions that determine how work is done. The former may have some salutary effect; some information about workplace hazards is undoubtedly better than none. One direction government can take now is to require, through legislation, that employers allow workers to have a strong voice on health and safety committees that have real powers—to inspect the workplace, to recommend policies, to hire independent health and safety professionals to advise them, and to shut down hazardous operations. Workers need more than the right to know, they need the right to act. And since unions have not won these protections for most employees in collective bargaining, the government must mandate it.

Government also needs to reorient its thinking about the nature of workplace protection. The current approach is too little too late. As is the case with environmental pollution control generally, whether clean-air or clean-water legislation, or the toxic waste laws, occupational safety and health regulation is reactive: Industry decides how to organize production, and then government acts, after the damage has been done, to clean up the consequences. This means that government is forever playing catch-up. And with new, untested chemicals continually introduced into the workplace, this is a no-win situation for government and workers. Effective workplace regulation requires that government change employers' basic approach to production. Our reliance on petrochemicals, for example, must be reexamined. Ultimately, this will mean a comprehensive, long-term process of conversion to another technology. The government is positioned to mandate this shift.

WHERE ARE WE NOW?

Worker protection is limited in the United States. The most encouraging developments are in science, where enormous gains have been made, especially in understanding the etiology of workplace disease. A growing network of public-health professionals now works with government and labor to improve workplace safety and health. This situation is likely to improve.

Other developments are less encouraging. Too many employers continue to resist regulation. OSHA inspectors routinely find high levels of noncompliance with agency standards, including widespread underreporting of accidents and injuries.[3] Declining profits, sluggish economic growth, heightened international competition—all lead employers to cut expenditures. And these cuts undermine health and safety by speeding up the assembly line; postponing equipment maintenance; shrinking work crews; hiring newer, inexperienced workers to replace higher-paid, more-experienced workers; and curtailing employee health and safety training.

Dramatically weakened in the last ten years, organized labor can no longer effectively watch over agencies like OSHA or mobilize workers to protect themselves on the shopfloor. A few unions, like the United Auto Workers, the Oil, Chemical, and Atomic Workers, and the United Steel Workers, continue to make this a priority issue. But the labor movement has been forced to refight struggles it thought long won, including the fight against the use of replacement workers to break strikes and undermine unions.

The wider political culture supports worker health and safety, but the issue does not appear as salient as other health and safety issues. Like the unions, most people seem most concerned about their economic survival. With unemployment and underemployment rates soaring, and health-care costs spiraling out of reach, workers look first for a secure job with benefits and only later think about the health and safety consequences of their job.

Finally, from 1981 to 1992, the federal government has been quite hostile to serious workplace regulation. Under Republican administrations, OSHA radically curtailed its standard-setting and enforcement activities. Because a hostile White House and Office of Management and Budget, more interested in protecting employers rather than workers, have forced regulators to meet nearly impossible burden-of-proof standards and cost-effectiveness tests, rule making on established carcinogens, such as cadmium and formaldehyde, have dragged on for years.

As a result, OSHA rules today cover only a small percentage of the toxic substances that threaten workers. The agency has set control levels for only 121 of the 526 toxic substances identified as carcinogens by the International Agency for Research on Cancer, the National Toxicology Program, (NTP) the American Council of Governmental Industrial Hygienists and the NTP/National Cancer Institute Cancer Bioassay Program.[4]

In the field, federal OSHA and the state-level agencies that have opted to implement the Occupational Safety and Health Act in state-plan states have been woefully understaffed, and, too often, politically constrained by pressures from employers and local political leaders more concerned about the business climate than workers' rights. Nowhere was this made more clear than in the fire that swept through the Imperial Food Products poultry-processing plant in North Carolina. Twenty-five workers died because a factory owner had illegally locked the exit doors to prevent pilfering. The agency charged with implementing the Occupational Safety and Health Act in North Carolina had so few inspectors on its payroll, and so little interest in regulation, that this plant had *never* been inspected, despite the knowledge that food processing is notoriously dangerous work.

THE FUTURE OF WORKPLACE REFORM

Although the analysis to this point is gloomy, it does not follow that workplace reformers should despair. Despite recent defeats, it is important to maintain a long-term perspective: Workplace protection has come a long way since the horrors of the nineteenth century, when workers were expected to live, or die, with workplace hazards. The common law assumed that, by taking a job, a person individually assumed the risks of that occupation. When government finally intervened in the late nineteenth and early twentieth centuries, it moved slowly and cautiously, establishing state-level factory inspector systems with little real power and state-level workers'-compensation systems that provided injured employees with minimal compensation. Workers were still assumed to be at fault in most instances, and employers retained tight control over the labor process.

From this perspective, substantial progress has been made in recent decades. Throughout the advanced industrialized world, governments have begun to change the way they think about workplace protection. Today's approaches are broader, more comprehensive, and more integrated. At least in principle, the links between workplace safety and health, public health, and environmental protection are readily acknowledged. The goals of state intervention have also changed from providing injured workers with what amounted to poor relief to forcing managers to rethink the impact of work on the labor force and society. Finally, regulation can work: When OSHA set and enforced strict standards for vinyl chloride, lead, and cotton dust, occupational exposure declined significantly.[5]

Of course, there are also some very disturbing trends. The institutional decline of the labor movement, the establishment of a North American free-trade zone, and the pervasive problem of unemployment all make it more likely that employers will ignore the social costs of production and that government will not protect workers. Struggles between labor and environmental groups over the fate of workers displaced by environmental protection also make it that much harder to mount a common campaign against corporate polluters. But there is nothing in this history to suggest that these obstacles cannot be overcome.

There are some promising trends. In Northern Europe, Canada, Austria, and several states in the United States, governments have experimented with greater worker involvement in the regulatory process through the establishment of worker committees with real powers. There is also growing international awareness of the importance of protecting worker health and safety, given the competitive pressures of an increasingly global market place. And coalitions between workers and environmentalists form more readily these days with greater mutual understanding.

NOTES

1. *Workplace Safety and Health,* (September 1990): 1.
2. See Barry Commoner, "After 20 Years: The Crisis of Environmental Regulation," *New Solutions* 1 (1990): 22-29.
3. The available research indicates that from one-third to one-half of the companies inspected still exceed OSHA standards for lead as well as silica, another commonly recognized hazard for which companies routinely test. See Charles Noble, "Keeping OSHA's Feet to the Fire," *Technology Review* (February/March 1992): 47.
4. Arthur Oleinick, William J. Fodor, and Marc M. Susselman, "Risk Management for Hazardous Chemicals: Adverse Health Consequences of Their Use and Limitations of Traditional Control Standards," *Journal of Legal Medicine* (March 1988): 57–58.
5. U.S. Office of Technology Assessment, *Preventing Illness and Injury in the Workplace* (Washington, D.C.: Government Printing Office,1985), 15.

18

Labor's Environmental Agenda in the New Corporate Climate

Eric Mann

Organized corporate power continues to circumvent traditional controls on its authority to affect the health and safety of workers and the communities in which they live through the control of production and resources. The mainstream Democratic Party, trade-union, civil-rights, and environmental leadership—all of whom represent large constituencies—have often taken an explicitly pro-corporate approach. While limited social controls on corporate power and other attempts to make the corporation accountable are important, what is truly needed is a grass-roots base that creates class-conscious environmental politics. Environmental justice is not merely a battle against pollution but a kind of politics that demands popular control of corporate decision making for workers and communities. The task is to bring together diverse constituencies based on a common vision.

In Europe, political parties are the organizational forms by which the conflicting interests and cultures of the oppressed can be mediated and focused into a general strategy. That is not true in the United States. We need to create, on a regional, primarily nonelectoral scale, laboratories for exploring new forms of direct democracy. This means addressing gender, race, and class contradictions in our own organizations; creating workable syntheses between environmental demands and the real employment needs of workers and communities; involving large numbers of people in direct confrontation with corporate polluters; and placing environmental activism in the context of rebuilding the progressive movement. What is needed is a long-term perspective that challenges institutional power and asserts democratic policy.

Before exploring our labor agenda, we need to consider why conventional approaches will not work. At one point early in the environmental debate, many believed that corporate executives and their children, having to breathe the air, eat the od, and drink the water, might feel a self-interested urgency about saving the planet. Several decades later, the cultural and ethical degeneracy of unmitigated free-enterprise capitalism—ideologically justified in concepts of deregulation, corporate competitiveness, cost effectiveness, and personal freedom—has produced a corporate elite that has shown itself thoroughly unable to grasp, let alone solve, the disastrous and at times irreversible effects of their production policies. At the same time, corporate executives declare themselves environmentalists and blame environmental destruction on all of us. Corporations often appoint token environmentalists to boards of directors, which are specifically charged with maximizing profits. Additionally, we have a new growth industry of toxic cleanup firms that rake in enormous profits from government Superfund contracts.

The new "environmental" corporate establishment manages to reduce both the production and cleanup of toxins to opportunities for profit and career, thus creating another layer of institutional control-inhibiting solutions. As more radical demands for the elimination of the production of toxins become widespread, corporations that profit from producing them and corporations that profit from cleaning them up will have a strong material interest in their continued existence. The institutional matrix is frightening: Corporate polluters derail environmental regulations in Congress; corporate pollution managers make lucrative deals that neither restrict polluters nor effectively clean up the toxins; government agencies set up ostensibly to protect the environment become captive to the polluters and pollution managers; and corporate boards of directors co-opt the most malleable and greedy environmentalists to clean up their image. Given this setup, grass-roots organizing must extend beyond populism to an analytical and strategic perspective.

WORKERS AND THE ENVIRONMENTAL MOVEMENT

Environmental activists who want to reach out to working people and their unions will find some encouraging places to start. An embattled but nonetheless substantial segment of today's labor movement pursues aggressive strategies of internal democracy, mass mobilization of the membership to reassert adversarial unionism, and a progressive political agenda that focuses on labor–community coalitions. These strategies find expression in a variety of movements ranging from the revitalized United Mine Workers and the union-reform movements of the Teamsters and the United Auto Workers (UAW) to the Service Employees' creative Justice for Janitors campaign and initiatives against U.S. intervention in the developing world.

The difficulties of involving working people in the leadership of a new environmental movement will center on the issue of jobs blackmail. Organizers must confront the ongoing dilemma of how to organize workers to oppose socially destructive corporate policies, such as employment discrimination and military production, that are not of their own making but provide them with a material livelihood. The economic and social insecurities of the past decade exacerbate workers' traditional fears of layoffs. With continuing cuts in employment, wages, health benefits, and social programs, and the attendant decay of working-class social and community institutions, jobs become the last remaining element of economic and social viability. Demands by environmentalists to shut down some of the most lethal corporate production processes will be perceived by workers as a direct threat to their well-being.

Too often the labor establishment not only refuses to provide an alternative institutional and ideological approach, it also exploits these fears to further reinforce the worker's perception that the only hope is to stick with management. Many major unions have abandoned an independent political program in favor of labor-management cooperation. This approach is hostile toward any movement that challenges management rights, and it pits the muscle of the union establishment against reformers who argue that labor should have some voice in the products and processes that management chooses.

A pro–working-class environmentalism, however, begins with the strategic understanding that many workers in plants that are poisoning the community are also themselves being poisoned in the workplace. In the past the occupational safety and health movement found itself trapped within the workplace, too often falling prey to labor-management cooperation, in which too much protest is bad for "competitiveness." But a movement that could synthesize the health and safety of the workers with community health has exciting organizing potential.

One proposal to address part of the problem of jobs blackmail is a superfund for workers. This would allow long-term training, education, and income maintenance for workers who have been laid off when a facility is shut down because of environmental concerns. The goal is to lessen the workers' ties to toxic production by offering a viable alternative.

In order for a demand such as this to have any substantial political impact, it must be offered not as a buy-off but rather as a part of a political worldview that assumes workers can understand the broader interests of their class and can respond to ethical as well as material appeals. A community-initiated environmental demand to stop production by an Exxon, Chevron, Monsanto, or Du Pont, combined with the presence of a superfund for workers, would reach workers not just as employees but as members of both the broader community and the *class* of workers. At the same time, there need to be far greater tax penalties for companies found guilty of polluting the environment, so that polluters pay most of the cost of toxic clean-up.

Additionally, workers should pressure companies for changes in their production processes. For example, any effective clean-air plan must involve a change in automobile design and energy source. But since ten million cars of any design clogging up the city will pose significant problems, a progressive solution must involve extensive, affordable public transportation followed by a dramatic reduction in auto use. On the surface, this would appear to contradict auto workers' interests. But the workers must take the lead in pressing auto companies to introduce new technologies and products. Electric cars may be one solution and a political opportunity for auto workers to be in the forefront of a movement for a less-polluting vehicle.

Another piece of the agenda is to restrict capital flight to deregulated states or to Mexico. When companies run away to avoid unions and regulation, they need to be penalized. Companies should therefore be compelled to sign at least ten-year commitments to remain in a community as part of any loan, tax incentive, or other government program to help business.

Workers must also oppose U.S. firms' dumping toxic waste in developing countries. Many international companies, when prevented from dumping waste products in U.S. communities of color, ship it to developing countries. This will require making connections and alliances with activists in Latin America, Asia, and Africa to discuss long-term strategies.

On the consumer side, corporations must be pressured to develop safe products. In this regard, the terms of debate need to shift from one that focuses on individuals purchasing environmentally safe products to collective demands that call for banning harmful products. This suggests the use of institutional consumer campaigns rooted in unions, community organizations, and religious organizations.

THE STRATEGIC IMPORTANCE OF COMMUNITIES OF COLOR

As reported elsewhere in this volume, corporations and governments site toxic waste facilities extensively in predominantly African, Latino, and Native American communities. Addressing the inequities requires challenging racial prejudice. In addition, given the decimation of the infrastructure of the inner cities—unemployment, deteriorating housing, low wages—we need new models of community economic development initiated by indigenous activists.

In East Los Angeles, for example, the furniture and metal-plating industries employ tens of thousands of workers in well-paying jobs but subject workers, along with the rest of the community, to extreme toxic exposures. In the service sector as well, janitorial, dry-cleaning, and auto-body shop workers are exposed to deadly chemical compounds. Residents of these communities often lack the resources to live elsewhere and have been least likely to express opposition. Thus it is vital that workers in factories, low-income people, and residents of communities of color make common cause.

The environmental movement too often thinks of workers and people of color as either stick-figure social categories or rare species, objectified at times through condescension. Each of these so-called categories represents a complex formation with its own social, material, and political differences.

Even those progressive environmentalists who do see the strategic importance of labor tend to equate the working class with people directly involved in the production of toxics. This misguided perspective gives disproportionate weight to one sector's specific demands for a job and not enough validity to the needs of the entire working class and of society as a whole.

CHALLENGING CORPORATE DECISION MAKING

The environmental crisis is about institutional and corporate production. Acid rain, global warming, pesticides, and toxic waste are connected with industries. But any efforts to limit or shape production in sound ways will involve direct confrontations between management's right to decide what a corporation will produce and the rights of workers and communities to live and work in safety. Strategies to build effective and democratic trade unions that could break with the current union pattern of slavish obedience to corporate priorities in return for short-term economic benefits for workers, as well as strategies to build city-wide and regional coalitions across race, class, and gender boundaries, become central to the creation of an effective environmental strategy that might limit the negative impact of corporate ecological policies. We will need new models for political and economic life, models that combine representative government at the top with significant power for direct input into decisions at the grass roots, both from workplaces and from communities affected by any given decision.

Yet this kind of democratic process will be resisted by corporations and will only be achieved if we have a transformation of labor unions as well as the development of powerful community coalitions. Oppositional movements in the labor movement, such as the Teamsters for a Democratic Union or the UAW's New Directions, need to integrate the legitimate demands for union democracy and better contracts with a more fundamental challenge to the toxic processes and products that characterize contemporary production in many firms. Ultimately, workers must demand a comprehensive program for a nontoxic economy and support worker retraining and income maintenance while moving the economy in an ecologically sound direction.

Community coalitions must be formed that develop regional economic plans from the bottom up that can work with the unions in a larger struggle for environmental justice. While an emphasis on transforming corporations may seem utopian, it is far more realistic than the present electoral and lobbying strategy that imagines effective regulation of toxins within present institutional arrangements.

Some steps toward developing a consciousness that might challenge management's rights have already been taken in various labor struggles in the last decades. In Los Angeles, dozens of groups organized a campaign in 1982 to keep General Motor's Van Nuys plants open. The campaign mobilized workers, Latino American and African American community members, and white progressives and threatened General Motors with a boycott of its products if the plant is ever closed.

After ten years of challenging General Motors to keep the plant open—with the unheard-of demand for a decade of work—the Van Nuys plant was finally closed in August 1992. For the Labor/Community Strategy Center, the tragic lesson is that an assertive workers' movement could be defeated and repressed by a joint effort of

General Motors' management and the UAW international union. This team concept broke the back of the local union and eventually paved the way for the plant's closing.

On the bright side, out of that struggle came the formation of the Strategy Center and ongoing work with labor groups throughout and beyond Los Angeles who are concerned about new technologies as part of a new unionism. Interestingly, the most encouraging relationship so far is with the Greenwork Alliance in Toronto, Canada, where a coalition of Canadian auto workers locals and Greenpeace Canada, in conjunction with the Strategy Center, is examining alternative production of environmentally sound products and public-transit vehicles to offer alternatives to workers in closed-down plants and in existing auto factories. "Green work, not pink slips" is their slogan, and the Labor/Strategy Center is starting to learn how to both build a social movement and to study and analyze industrial production and processes.

The logic of these struggles leads beyond attacks on specific corporate offenders to a need for regional strategies that necessarily raise more fundamental distribution questions. Though many workers have substituted the shopping mall for the union hall as the center of their recreational and cultural life, there is growing awareness that rampant materialism can offer little real satisfaction or sense of meaning and purpose.

A movement that sought to conserve the environment by reducing the quantity of goods produced, while simultaneously advocating more egalitarian distribution of goods, could gain the allegiance of many working people in the years ahead. The political right's vision of unchecked corporate action and a state sector designed primarily to serve corporate interests may be increasingly vulnerable to ecological challenges, particularly if an ecologically based campaign were linked to plans for a strong safety net for the unemployed and the poor, guaranteed medical and health care, low-cost and high-quality public education and transportation systems, and the use of tax revenues for the support of new cultural endeavors.

CONCLUSION

A deepening ecological crisis requires that we move beyond narrow and supposedly more realistic approaches to objectives that can address the full depth of the crisis. This will involve a more rational planning of production of resources. Yet only a powerful grass-roots movement can plausibly develop the strength to counter those corporate interests that will continue to oppose rational planning to meet human need. After almost two decades of bipartisan eulogizing of the civilizing role of market forces, our political, material, and ethical environments are deteriorating rapidly. The ecological crisis requires that we abandon self-interested, single-issue struggles and produce a visionary and radical movement. In part, this means connecting environmental discussions to nonpolluting economic development, rebuilding the inner cities of the nation, and not separating workers from community issues.

The environmental crisis is not solvable locally. Thus, while grass-roots movements are essential building blocks and catalysts, they cannot be substituted for a broader political strategy to transform policy and power at the national level.

During the 1980s, the retreat of many movement organizers into single-issue specialization was partially a product of a loss of confidence in reform, social democracy, or any broader worldview that could give coherence and optimism for social transformation. Today, however, as the result of the futile efforts to regulate increasingly concentrated corporate capital, movement survivors of the 1980s seem again willing to explore macro-strategies and agendas for economic and political democracy.

A recent example from the work of the Strategy Center may be illustrative. For the past two years, it has been building a very successful community-based movement to implement a community right-to-know law, led by Strategy Center organizers in Wilmington, East Los Angeles, and South Central Los Angeles, along with our close allies, Mothers of East Los Angeles (MELA) and Concerned Citizens of South Central Los Angeles. But in April 1992, we were challenged by the introduction of a bill in the California legislature, initiated by the Western States Petroleum Association, that would have removed the authority of regional air districts to set standards higher than the federal EPA. Because Los Angeles has the worst air quality in the United States, and the Quayle Commission on Competitiveness had been working to gut federal air quality standards at the EPA levels, the passage of this bill would have undermined the right-to-know law that had been the statutory basis for our organizing work.

As progressives again debate the merits of a radically reformed capitalism vs. new models of democratic socialism, the content of economic democracy must center on the replacement of our present model, in which private corporate power dominates public life, with new models of public power and decision making rooted in workplaces and communities. Those supporting environmental justice need an infusion of courage and vigor to break with tepid reformism and corporatist co-optation, and to confront the logic of the problem they have posed. This will require a challenge to economic rules and power relationships.

19

Corporate Plundering of Third-World Resources

Robert Weissman

SINCE AT LEAST the time of Columbus and the conquistadors, rich, powerful societies have been stealing resources from poorer and militarily weaker ones on a global scale. Global plundering has always had devastating environmental, as well as social, effects. By nature, resource extraction degrades the environment; when executed with little or no concern for its ecological impact, the results are usually far-reaching and tragic.

After World War II the world's richest nation, the United States, continued the pattern set by the European colonizers. With multinational corporations as the agents of expropriation, modern resource extraction, and transportation technologies, the transfer of the Third World's natural resources to the industrialized countries accelerated to an unprecedented pace, causing unprecedented environmental destruction.

During the Pax Americana, military power has continued to play a central role in nations' ability to control resources in other countries, evidenced most recently in the Persian Gulf war.

However, in the 1980s and into the 1990s, military power has receded as the key means to ensure multinational corporate access to Third-World resources. The new favorite club of industrialized country power brokers is the massive foreign debt of Third-World countries.

CAUGHT BETWEEN AN INVISIBLE AND A VISIBLE HAND

In the 1970s, flush with deposits from the increasingly wealthy OPEC countries, the world's major commercial banks faced the challenge of "recycling" deposits—that is, loaning them out so that the banks would be able to pay interest on the deposits and still earn profits. A central element of the bankers' eventual solution to this dilemma was to loan huge sums to Third-World countries. The widely held theory was that since countries do not go bankrupt, the banks would be able to collect interest and principal repayments for their loans to the Third World, even if the loans went to unsound projects without a sufficient return on investment.

When interest rates skyrocketed in the late 1970s and early 1980s, the interest charges on the Third-World debt went up in tandem. That led to the ballooning of the debt, as countries borrowed extensively just to pay back old debts, and created what came to be known as the Third-World debt crisis. The total Third World debt stood at slightly more than $500 billion in 1980, reached $965 billion by 1985, and grew to almost $1.3 trillion by the end of the decade. According to the World Bank's definitions, forty-one developing countries, including Argentina, Mexico, Kenya and Nigeria, are severely indebted, and twenty-eight others, including India and the Philippines, are moderately indebted.[1]

Heavy debt burdens place Third-World countries in an inescapable bind. To earn money to pay back interest on their debt, they must increase their export revenues. And their natural-resource base is the easiest source of export revenues. So Third-World governments—from Brazil to Bangladesh to Cameroon—are forced to mine minerals, harvest trees, or drill for oil at even higher rates for an even more minimal return.

The debt crisis, and its consequences, is the working of the "invisible hand" of the marketplace. The complementary "visible hand" belongs to the major multilateral lending agencies, the International Monetary Fund (IMF) and the World Bank. Through imposed "structural adjustment" programs, these entities are able to exert almost irresistible pressure on debtor countries to expand resource exports.

When Third-World debtor nations find themselves without the money to repay their commercial bank loans, the primary solution available is to borrow more. But private and governmental lenders alike are unwilling to provide funds to debtor nations which do not receive the IMF stamp of approval. Since most Third-World countries maintain huge debts and need new loans to avoid default, the IMF exerts enormous influence over these countries' economic and social policies, even though it does not provide huge loans itself.

The price of accepting an IMF loan or many of those made by the World Bank is an agreement to enact austerity and economic liberalization measures, together known as a structural adjustment program. The measures range widely, from trade liberalization and currency devaluation to interest rates hikes to cuts in government spending and privatization of state-owned enterprises. By freeing up market forces and correcting distortions in the economy, the IMF and World Bank expect poor countries to increase export earnings and cut expenditures so that they can reduce their balance-of-payments deficits.

The operations of the multilateral lending agencies truly touch every corner of the Third World. In fiscal year 1992, the IMF had structural and enhanced structural adjustment arrangements in effect with fifty-four countries; in the same year the World Bank made structural adjustment loans to twenty-two countries.[2]

The social effects of these programs are tragic: Slashed budgets for health care and sanitation contribute to outbreaks of cholera; cuts in government spending throw people out of work in countries without social safety nets; devalued currencies raise the cost of imports, including the costs of vital needs like food and medicine.

The environmental effects are equally devastating. Promoting foreign investment and orienting Third-World economies toward exports, when they have almost nothing to offer besides cheap labor and natural resources, inevitably exacerbates the pressures on delicate Third-World ecosystems, which house resource riches.

THE DYNAMIC OF DEBT AND DESTRUCTION

Together, the invisible hand of the market and the visible hand of the IMF and World Bank maintain a tight grip on Third-World debtor nations. Practically without fail, these countries loosen or remove restrictions on foreign investment, abandon efforts to develop an economy oriented to meeting local needs, and strive desperately to increase their exports.

In a rush to lay claim to valuable resources, foreign companies destroy the local environment and endanger the cultural and often physical survival of the indigenous people who populate it. The story is depressingly similar in countries all over the globe, with only a different set of corporate, government, and indigenous actors.

Indonesia

In Indonesia, where widespread poverty exists in a land of immense natural-resource wealth, a ruthless partnership of foreign investors and domestic elites have taken advantage of the country's debt-driven economic liberalization to make millions at the expense of the country's indigenous population and environment.[3]

Indonesia's foreign debt totals nearly $40 billion. The country has received huge infusions of cash from the World Bank—$1.2 billion a year in the second half of the 1980s—for both structural adjustment and sector-specific loans.

The Indonesian government's economic liberalization campaign began in 1983 and intensified in 1986. With the rate of economic growth and the price of the country's main export, oil, declining, the government abandoned its old model of development and its commitment to protecting national producers. In place of import substitution, it began pursuing a strategy of export-led growth. It devalued its currency (the rupiah) 28 percent, adopted an austerity budget, liberalized trade, and substantially deregulated the financial sector.

Each year, more than one million hectares of Indonesia's tropical forests, one of the richest genetic storehouses in the world, are destroyed, in significant part to earn currency to pay off the country's debt. Indonesian President Suharto articulated the relationship between Indonesia's debt and its deforestation in the 1970s, when deforestation rates were only one-third of current levels: "We do not have to worry our heads about debts, for we still have forests to repay those debts."[4] Wood products, Indonesia's main non–oil and gas export earner, now bring in more than $3 billion annually. Indonesia produces about 70 percent of the world's hardwood supply.

Corporations are contributing to the irreversible destruction of Indonesia's forests. They routinely ignore the country's few environmental regulations governing logging, cut protected trees, fail to replant, and exacerbate the damage they cause by road-building and other infrastructure-support activities. Destruction of forests involves not only cutting trees but endangering Indonesia's incredibly varied plant and animal life, including a large number of species unique to the country.

Logging operations also threaten many of Indonesia's three million indigenous people. In the northern tip of Sumatra, for example, intensive logging has caused unprecedented flooding, destroying rice fields and threatening thousands with starvation. In West Papua, the Moi people have seen the forests that provide them with fruits and vegetables and meat decimated by timber companies.[5]

Papua, New Guinea

Suffering from a crippling debt of $2.5 billion, Papua New Guinea has opened up its rich natural-resource base to foreign investors. Multinational logging, mining and now oil companies have entered the country and operate with little restraint.[6]

Despite the 1990 implementation of a two-year ban on new logging permits, logging on Papua New Guinea is accelerating, mostly by Japanese companies, which comprise 90 percent of the country's timber industry. Companies like Japan and New Guinea Timbers (JANT), a wholly owned subsidiary of the Japanese Honshu Paper company, are clear-cutting for pulp. Corporate practices are not significantly different than they were in 1987, when a government-commissioned inquiry into logging practices found "that some companies [roam] the countryside with the self-assurance of robber barons; bribing politicians and leaders, creating social disharmony and ignoring laws in order to gain access to, rip out and export the last remnants of the province's timber."[7]

Mining has also taken a huge toll on the country's environment. From 1972 to 1989, CRA, a subsidiary of the British mining giant Rio Tinto Zinc, operated the Panguna copper and gold mine on Bougainville Island, a coastal island five hundred miles northeast of Papua New Guinea. CRA discharged mine tailings directly into the Jaba and Kawerong Rivers, killing most aquatic life.

After local landowners had waited a year for CRA to respond to their 1988 complaints about environmental and social effects of the mine, as well as the low royalties they received, mine employee Francis Ona formed the Bougainville Revolutionary Army (BRA). The BRA waged a bloody sabotage campaign against the mine, leading CRA to close the mine at the end of 1989. The Papua New Guinea government responded harshly to the BRA, since the Panguna closing has deprived the country of 40 percent of its export revenues, revenues critically important because of the country's debt-repayment obligations.

In 1989 and 1990, Papua New Guinea military forces destroyed thousands of Bougainville village homes and committed widespread human-rights violations. Papua New Guinea maintains a total embargo on all medical and other supplies destined for Bougainville, resulting in the deaths of thousands on the island, according to Bougainville representatives. Papua New Guinea maintains military control over the northern tip of Bougainville.

Another mining company, Ok Tedi Mining Limited (OTML) has operated a mine near the West Papua border with the same reckless abandon as CRA. Jointly owned by U.S.-based Amoco (30 percent), the Australian Broken Hill Proprietors (30 percent), the Papua New Guinea government (20 percent) and a German consortium, OTML dumps 80,000 tons of sediment into the Ok Tedi River each day. Periodic flooding has deposited sediment load contaminated with heavy metals onto the river banks, making them unsuitable for farming. OTML has twice spilled large quantities of sodium cyanide into the water.

Landowners near the Ok Tedi mine have also protested the effects of its operations, closing it several times. But OTML is increasing its output of copper and silver at the urging of the Papua New Guinea government, which is still struggling to make up for losses at Bougainville.

The Papua New Guinea government is also encouraging Chevron to undertake the giant Kutubu oil project, expected to produce 170 million barrels of oil over a ten-year period. The project includes construction of a 163-mile-long pipeline (which extends through one of the largest expanses of mangroves in the world), a tanker-loading buoy, a mile-long airstrip, a sixty-five–mile road, and a network of roads from the well sites to the central processing facility. Environmentalists fear the project could bring new environmental catastrophes to Papua New Guinea.

Guyana

Remote areas of the Amazonian region of Guyana, long the home of indigenous populations and undisturbed by other people, are besieged by multinational corporations, the beneficiaries of the country's tremendous debt burden.[8]

At the end of 1988, Guyana, a country with a per capita gross national product of only $350, owed $1.7 billion to foreign lenders, more than $2,000 for each of its under 750,000 people.

The country has been subjected to an IMF–World Bank debt-rescheduling program. In an attempt to check spiraling inflation rates, the multilateral lending institutions forced Guyana in March 1991, to devalue its currency by 56 percent. To boost foreign-exchange revenue, the IMF and World Bank are pushing the country to increase exports of its interior natural resources: timber, other forest products, and minerals.

Until 1990, Guyana logged only a tiny portion of its extensive forests, but debt-related pressures are leading the government to grant large concessions to a Venezuelan company, Palmaven, to a South Korean–Malaysian consortium, Sung Kyong, and to other foreign corporations. A road designed to open up Guyana's interior, financed by the Brazilian external-financing agency, has crossed indigenous lands and even cultivated fields; residents have not been compensated. Community leaders fear that the road will lead to the eventual destruction of the Makushi people.

The government is also encouraging foreign companies to exploit the country's extensive mineral deposits, especially gold, despite the severe health hazards of gold mining. The cyanide used by commercial mines, run by companies like Canadian Golden Star Resources, to isolate gold from alluvia and the poisonous tailings generated from crushed rock seriously threaten water quality and fish stocks. New "missile dredges"—huge remote-controlled vacuum cleaners that pump water into the alluvial deposits and suck them up to process out the minerals—dig deep into the river banks, sometimes as far as seventy meters, and liquify the mud and gravel. Indigenous communities report that silting has polluted water supplies and reduced fishing returns as far as 60 kilometers downstream.

Guyana's sparsely inhabited interior is the traditional preserve of a number of indigenous groups, such as the Karina (Carib), the Pemon (Arekuna), the Kapon (Akawaio), and the Wai Wai, and they are suffering the most from Guyana's debt-driven crash-development program. Because of policies designed to pay off a debt they had no hand in acquiring, they are experiencing the takeover of their lands, a rising tide of malaria and numerous diseases brought in by miners, the pollution of their rivers, and the decimation of the fish and game upon which their livelihoods depend.

STRATEGIES OF RESISTANCE

Short of following the Bougainville lead and forming a guerrilla army to expel multinational corporate investors, what strategies can local Third-World groups and their allies in the North pursue to prevent multinational corporations from plundering Third-World resources and despoiling the Third-World environment?

Targeting specific corporate investment proposals

One strategy targets specific corporate investment proposals. Coalitions of Third-World environmental organizations, indigenous and community organizations likely to be affected by a particular project, join with activist organizations in industrialized countries, especially the corporations' home countries, to challenge logging or mining projects. They hold protests, take out advertisements in newspapers, organize letter-writing campaigns, and threaten to initiate consumer boycotts.

This strategy has achieved some successes in the last several years:

- In 1989, the Scott Paper Company pulled out of a proposed $650 million eucalyptus tree plantation and pulp mill project in the tropical forests of West Papua due to international grass-roots pressure. The project involved tearing down existing forests and planting eucalyptus trees in large parts of a 500,000 hectare lease from the Indonesian government. Eucalyptus is a desirable tree species for pulp and paper companies because it grows quickly, but it is particularly harmful to soil and the surrounding environment because it drains the soil of nutrients at a high rate and requires a huge amount of water. Local indigenous groups and their supporters claimed that the eucalyptus plantation and supporting infrastructure would destroy the land and resources on which the 15,000 indigenous people in the region rely.
- In 1991, Conoco Oil, a subsidiary of Du Pont, announced it would not follow through on plans to build ninety miles of roads into pristine Ecuadoran forest in order to pump oil from its vast concession in the territory of the Huaorani people. Again, a coalition of local indigenous and environmental groups and international supporters stopped the project.
- In early 1992, the Stone Container Company sought to lease Honduras' remaining virgin pine forests. The contract Stone's lawyers drafted would have given the company virtual impunity in harvesting 900,000 acres of pine forest in eastern Honduras. Because the contract was so disadvantageous, Honduran environmentalists and Miskito indigenous groups were joined in their opposition to the proposals by other social sectors, including business people. With some additional help from environmental groups in the United States, the Honduran coalition succeeded in pressuring the government to reject the contract.

These three cases are an impressive testament to the ability of well-organized citizens, working cooperatively across borders, to influence major investment decisions. Such efforts are absolutely necessary to respond to specific proposals.

However, as the participants in these strategies usually realize, targeting specific proposals can only be a stop-gap measure. Citizen groups lack the resources to monitor and challenge every potentially environmentally destructive project, and even success in stopping a particular proposal is often only temporary. Indonesian activists now worry that a Japanese company will take up Scott's Irian Jaya project; the Dallasbased Maxus Energy Corporation has taken a controlling interest in what was formerly Conoco's Ecuadoran prospecting area; and the Stone Container Company, after undertaking a public-relations campaign in the United States and Honduras, has reintroduced its proposal, with a few minor modifications, to log Honduras's pine forests.

Debt-for-Nature Swaps

The clear lesson from campaigns protesting specific multinational proposals is that stopping specific proposals is inadequate; land and resources must be protected. One method gaining in favor among mainstream environmentalists is debt-for-nature swaps.

Debt-for-nature swaps involve Northern environmental groups paying off part of the debt of a Third-World country in exchange for the country agreeing to protect forest land. They have been pursued most vigorously by the World Wildlife Fund (WWF) and The Nature Conservancy.

In theory, all parties benefit from the debt-for-nature arrangement. Banks have devalued loans repaid before their value sinks even further; debtor countries' debt burdens are lessened; and the environment is protected.

In practice, debt-for-nature swaps do not work so well. The amount of debt paid off is so small in relation to the overall debt owed by many Third-World countries that the payoff is virtually irrelevant.[9] And the protected lands, managed by local conservation groups, are often not protected. Independent reports on areas set aside in debt-for-nature swaps in Ecuador and Bolivia, for example, emphasize that they have been despoiled by companies seeking to exploit the natural-resource value of the land.

The most significant problem with debt-for-nature swaps, however, is how they relate to indigenous people. An unstated premise of the swaps is that the set-aside land is the government's to set aside. But indigenous people who live in these areas—though they may not possess legal title to the land—stridently reject this notion, arguing that the land is theirs. And they emphatically reject the idea that they should bear the burden of paying off a debt they did not contract.

Many progressive Third-World political and environmental activists also oppose any steps that legitimize the debt. In Brazil, groups tied to the Workers' Party of Luiz Ignacio "Lula" da Silva have protested the efforts of the WWF to put together a debt-for-nature swap. They charge that the debt was incurred by a military dictatorship to which banks should not have made loans, and that it should not be repaid at all.

A Comprehensive Approach Emphasizing the Rights of Indigenous Peoples

As well as violating the sovereignty of indigenous peoples, debt-for-nature swaps are based on a fundamentally flawed approach to protecting pristine ecosystems. Setting aside land is a necessary but wholly insufficient solution, since it fails to address the root causes of degradation of Third-World environments. The Latin American and Asian experiences with "protected" reserves clearly indicates that such areas will be violated if social pressures continue.

Efforts to protect Third-World ecosystems from multinational corporate plunderers should emphasize the following three principles:

First, indigenous people and forest-dwellers should be guaranteed custody of the lands they occupy in order to preserve precious ecosystems for future generations. They are morally entitled to those lands, and all experience indicates that indigenous people have the skills and priorities necessary to preserve them. They protect forests and other delicate ecosystems because those places are their homes, provide their sustenance, and are integral to their culture.

Organizations like the London-based Survival International and the Penang, Malaysia-based World Rainforest Movement have established the ties needed to work directly with indigenous people. And the indigenous themselves are increasingly well organized, within countries and internationally. In February 1992, at a conference held in Penang, indigenous representatives from forest-dwelling communities in the Americas, Asia, and Africa formed the International Alliance of the Indigenous-Tribal Peoples of the Tropical Forests. A charter adopted by the new alliance asserts that "there can be no rational or sustainable development of the forests and of our peoples until our fundamental rights as peoples are respected."[10] The charter advocates a new development approach: in place of large-scale development projects, such as logging and mining, it calls for small community initiatives under the control of the people who live in the forest. Indigenous people made additional progress a few months later at two major meetings held at the United Nations Conference on Environment and Development in Rio de Janeiro in June 1992.

Second, the external pressures that encourage Third-World countries to exploit their resources, primarily those pressures brought on by their foreign debts, must be alleviated. As long as foreign countries are forced to organize their economies toward earning foreign exchange to meet outrageous debt-repayment schedules, they will have almost no choice but to pillage their resource bases. Citizen activists in industrialized countries should work to encourage their governments to forgive the debt, in whole or in large part.

They should also work to pressure the IMF and the World Bank to abandon their fierce adherence to structural adjustment policies. Though the IMF and the World Bank are not accountable to any citizen constituency, they do receive their funding from industrialized countries, and citizens of those nations are uniquely positioned to exert some leverage on the brutal bankers. By calling for a withholding of funds or proposing that funding be made contingent on changes in the IMF and the World Bank's policies, citizens can make these agencies at least take into account the social and environmental costs of the policy measures they advocate.

A coming threat equally significant to the debt and structural adjustment programs is the General Agreement on Tariffs and Trade (GATT), the agreement governing most world trade. Proposals made in the current round of GATT negotiations—in many cases written by multinational corporations—would lock in proforeign-investment policies in Third-World countries. Because GATT negotiators have not yet agreed on final proposals relating to foreign investment and other important issues, there is an opportunity for citizens, through their national governments, to head off an incredibly far-reaching threat to the health of the planet's people and environment.

Third, a measure of corporate accountability must be achieved. This is the most distant goal. In the short- and medium-term, citizens can highlight specific corporate abuses (e.g., the Exxon *Valdez* oil spill had an enormously important educational effect); support boycotts against especially corrupt and irresponsible corporations; and support efforts to establish minimal corporate standards of conduct, such as the proposed—and very weak—United Nations Code of Conduct for Transnational Corporations. The no-longer existent UN Center for Transnational Corporations proposed developing minimal international corporate environmental responsibilities similar to the international human rights standards applied to governments. No one even discussed this proposal at the official conference, but citizens can work to put it on the United Nation's agenda.

In the long-term, multinational corporate activity should be strictly regulated, with nations given the right and power to reject foreign investment in favor of more controllable domestic, especially small, community-based enterprises. All major corporate investment should be preceded by publicly accessible social and environmental impact assessments, and all multinational corporate activity should be held to a best-practices standard. If corporate practices are not socially and ecologically sustainable, they should not be permitted. Criminal sanctions, with stiff penalties designed to make it unprofitable to violate environmental and social regulations, should await corporate violators. And victims of multinational corporations' activities should have access to justice in the corporations' country of origin.

These steps will not be enough to protect either the environment in the Third World or the interests of indigenous peoples. Domestic pressures on pristine ecosystems—from colonists, ranchers, or miners—must be relieved by land reform above all.

Still, it is a virtual certainty that, without the abatement of international pressures, many intact Third-World ecosystems will be destroyed, with many indigenous populations and much of the world's biodiversity decimated in the process. Northern citizens have an obligation to indigenous people, to people in the Third World, to the planet, and even to their own long-term self-interest to rein in U.S. and other Northern corporate plunderers and to change the conditions that give those companies so much leverage over the poor and weak countries of the world.

NOTES

1. *World Debt Tables 1991–1992: External Debt of Developing, Countries,* vol. 1 (Washington, D.C.: World Bank, 1991), 116–120.
2. *The World Bank: Annual Report 1992* (Washington, D.C.: World Bank, 1992), 20.
3. See Peter Halesworth, "Plundering Indonesia's Rainforests," *Multinational Monitor* (October 1990): 8–11; and Robert Weissman, "Rich Land, Poor People: The Economics of Indonesian Developments," *Multinational Monitor* (October 1990):16–21.

4. Quoted in "Systematic Destruction of Asia's Amazon," reprinted in the *Endangered Rainforests and the Fight for Survival* (Penang, Malaysia: World Rainforest Movement, 1992), 205.
5. See Anna Laine, "Moi Resistance: Standing Up to Indonesia," *Multinational Monitor* (October 1992): 12–14.
6. See David Minkow and Colleen-Murphy Dunning, "Assault on Papua New Guinea," *Multinational Monitor* (September 1991).
7. Ibid.
8. See Marcus Colchester, "Sacking Guyana, "*Multinational Monitor* (September 1991).
9. See Rhona Mahony, "Debt-for-Nature Swaps: Who Really Benefits?" *Ecologist* (May/June 1992): 97–103.
10. Marcus Colchester, "Indigenous Unite," *Multinational Monitor (*April 1992):6.

20

Global Economic Counterrevolution

The Dynamics of Impoverishment and Marginalization

Walden Bello

MOST PEOPLE IN the Southern Hemisphere remember the 1980s not as the decade of triumphant capitalism but as the decade of reversal. By 1990, per capita income in Africa was down to its level at the time many African countries achieved their independence in the 1960s. And in Latin America, per capita income in 1990 had not exceeded its 1980 level.

While people in the South empathized with the newfound freedoms of the citizens of Eastern Europe and the former Soviet Union, many of them could not, however, understand the rush of the postsocialism leaders to embrace radical free-market reforms and the International Monetary Fund (IMF). In their view, these were the very methods that had triggered a massive economic debacle in the Third World in the 1980s.

Socialism was undermined largely by its internal failings, particularly its inability to institutionalize democracy and its failure to create an economy that promoted equity without stifling growth and innovation and destroying the environment. The rough equality in living conditions under centralized socialist rule was not the dynamic equality amidst rising living standards and growing freedom envisioned by

---

The author would like to thank Kira Kim for research assistance and acknowledge the reliance on Priscilla Enriquéz for research on hunger in America, part of which resulted in the Food First Action Alert entitled "An Un-American Tragedy: Hunger and Economic Policy in the Reagan-Bush Era" (Summer 1992). The author, however, takes full responsibility for the facts and analysis in this paper, adapted from a longer article.

the pioneers of socialism but the equitable sharing of shoddy material goods and services amidst generalized economic stagnation and political repression.

The eagerness for free-market reforms on the part of Eastern Europeans failed to excite not only people in the Third World but many in the United States as well. Many Americans linked the free market and the unrestrained freedom of corporate capitalism to the sharp reversal of trends in U.S. living standards during the twelve-year Republican administration.[1] The same forces were likely producing similar trends in both the North and the South. The reversal of people's fortunes in both the South and the United States is better understood as a global counterrevolution—the Reagan-Bush Counterrevolution. The rolling back in the 1980s of the gains made by the Third World in the 1960s and 1970s, when U.S. foreign policy was dominated by liberal containment strategy, was not unrelated to the reversal of the concessions to American labor that had been institutionalized in the New Deal state constructed by Franklin D. Roosevelt and his successors.

RESUBORDINATING THE THIRD WORLD

The Insurgent South

The 1960s and 1970s were years of significant gains for the Third World, or South. On the one hand, national independence movements came to power or were institutionalized in Cuba, Mozambique, Angola, Guinea-Bissau, Vietnam, Laos, Kampuchea, Nicaragua, Iran, and Zimbabwe. On the other hand, stimulated by development strategies in which the state played a central role, Southern economies grew, with Latin America's gross domestic product (GDP) rising by an average of 5 percent per annum between 1960 and 1982, and Africa's by an average of 4 percent. The Asia-Pacific region, with its much touted "tiger economies"— Taiwan, South Korea, Hong Kong, and Singapore—had regional GDP registering an average increase of 7 percent per year.[2] A Third-World consciousness emerged from these developments, a sense of sharing similar interests vis-à-vis the North, was articulated into an agenda (the "New International Economic Order" or NIEO) adopted by the United Nations Special Session in 1974.

The rhetoric of solidarity seemed on the verge of becoming reality in the early 1970s, when the Organization of Petroleum Exporting Countries (OPEC) managed finally to seize control of the price of oil. OPEC's success triggered similar attempts by Third-World countries to create cartels in other raw materials and in agricultural commodities. When Third-World governments flocked to Paris in 1975 to confront the North at the crucial Conference on International Economic Cooperation (CIEC), many expected that the OPEC producers would unite with them to demand a comprehensive deal on a wide range of commodities.

The Demise of the Liberal Containment Strategy

The rise of the Third World coincided with the hegemony of anticommunist Cold War liberals. They also believed that growing Third-World markets were in the interest of U.S. capital and that to be successful, anticommunist insurgency had to be accompanied by a degree of economic prosperity.

The liberal approach was evident in the circumstances surrounding the creation of the International Development Association (IDA), the World Bank's "soft-loan" arm, in 1960. IDA was seen as a key instrument in the liberal strategy of limited redistribution of global wealth, but making it an affiliate of the World Bank ensured that the loan process would be controlled by the United States. It was designed to contain not only communist movements but also Third-World nationalists demanding fundamental changes in North-South relations.

But these policies did not prevent U.S. troops from going into combat in Vietnam in the early 1960s. More than a military expedition, Vietnam was also an experiment in the liberal strategy of containment and an attempt to engineer a capitalist revolution by promoting democratic reform of a corrupt feudal system of rule, urging land reform, and pouring in aid.[3] U.S. conservatives, who favored a straight-out military solution with no pretenses at reform, thus saw Vietnam as the failure of the liberal formula for containment, and they were doubly incensed by what they saw as Vietnam's "demonstration effect" on the rest of the Third World. For them, there was a linear progression from Vietnam's successful defiance of the United States to the Arab oil embargo in 1973 and the declaration of the NIEO in 1974.

OPEC raised oil prices substantially a second time in 1979. But even though OPEC had gone back on its promise to use oil as a weapon to get a whole range of concessions for the South during the CIEC negotiations in 1975, the specter of a unified Southern bloc controlling strategic commodities haunted the North.

The Reagan Project

The Reagan administration's agenda was to discipline the Third World. More lethal in its consequences than United States military adventures against radical Third-World movements and governments was the global economic warfare that the U.S. unleashed against the South. While the social interests and ideological predisposition of the Reagan coalition defined its broad strategic perspective, the formulation and implementation of its concrete policies evolved out of changes in the international balance of forces and relationships among the factions within the reigning coalition.

International economic conditions in the early 1980s, for example, favored those within the administration who sought an offensive, confrontational strategy toward the South:

- OPEC failed to deliver on its promise to use oil as a weapon for the larger cause of Southern solidarity. This was the product of a skillful maneuver to detach the OPEC countries from the rest of the Third World.[4]
- The prices of raw materials dropped to their lowest level since the 1930s. With many Third-World countries dependent on one or two commodities to gain foreign exchange, the drop in prices meant a severely constrained ability to import industrial goods, pay for imported food, and service their mounting foreign debt.
- The flow of multinational investment southwards in search of low-cost labor was threatened by the adoption of flexible automation methods, which promised to make high-tech production in advanced countries cheaper and more efficient than labor-intensive production in the Third World.
- But perhaps the Third World's most vulnerable point was piling up $700 million worth of debts to U.S. and other Western banks, which had competed intensely with one another to make loans to Southern governments in order to make profits from the huge sums of OPEC money deposited with them.

Conservative thinkers identified strategies of nationalist development via protectionism and strong state leadership as the key obstacles to the free flow of U.S. capital and exports. Moreover, aid was thought of mainly as a political instrument to achieve strategic purposes, and only secondarily for development. The conservative economists challenged the foundation of the liberal approach to the Third World: the idea that a prosperous capitalist Third World was in the U.S. interest. Third World growth, in their view, should be promoted by unfettered free enterprise, not by income redistribution schemes.

The task was to develop firm policies aimed at resubordinating the increasingly unmanageable Third World within the U.S.–dominated world economic system. The World Bank became a principal battleground between liberal and Reaganite approaches to the Third World. While some of the more ideological right-wing thinkers attacked the Bank as favoring state-dominated economic systems in the Third World and as a development bureaucracy that doled out $12 billion in welfare checks, pragmatists in the Reagan administration came to view the Bank as a useful, if not central, instrument in their effort to discipline the Third World.

The first salvo in this campaign was cutting the U.S.–promised contribution to the 1982 IDA replenishment by $300 million, leading other Western countries to cut their own contributions. Then, the United States pushed the World Bank to shift more of its resources from traditional project lending to "structural adjustment" lending. Structural adjustment loans (SALs) were systematically used by the Reagan Department of the Treasury to force open Third-World economies. To receive SALs, governments were required to lower barriers to imports, remove restrictions on foreign investment, reduce the state's role in the economy, reduce spending for social welfare, and make export production more attractive by cutting wages and devaluing the local currency. While most World Bank economists tried to sell these measures as necessary to promote efficiency, Third-World leaders accurately perceived them as striking at the heart of the Southern project of gaining more economic independence at the national level and effecting income redistribution at the global level. Few countries lined up to receive SALs initially.

Debt Crisis and Structural Adjustment

With the eruption of the debt crisis in 1982, more Third-World countries could not service the huge loans made to them in the 1970s. The Northern banks often made the adoption of a World Bank structural adjustment program essential to debt rescheduling, arguing that the structural reforms ensured that the debtors would be able to service their debts beyond the short term. Unable to persuade private banks to agree to reschedule and refinance their debts without the World Bank seal of approval, twelve of the fifteen debtors—including Brazil, Mexico, Argentina, and the Philippines—had submitted to structural adjustment programs by the end of 1985.

By 1991, lending for structural adjustment rose to 25 percent of total Bank lending. Whereas in the previous division of labor between the two institutions the World Bank was supposed to promote growth and the IMF was expected to monitor financial restraint, their roles now became indistinguishable as both became enforcers of the North's economic rollback strategy.

Structural adjustment programs have functioned as a terribly efficient debt collection mechanism. The net flow of financial resources to the Third World has been negative since the early 1980s. While U.S. commercial banks with huge claims on Third-World resources were the main recipients of this transfer, the World Bank and the IMF, which had originally been designed to facilitate a net transfer of aid resources to the Third World, were also beneficiaries.[5]

More consequential, however, than debt collection and the massive redistribution of financial resources from the South to the North was the attainment of the strategic objective of the Reagan Counterrevolution: the imposition of "reforms." From Argentina to Ghana, state intervention in the economy has been drastically curtailed; protectionist barriers to Northern imports have been comprehensively eliminated; restrictions on foreign investment have been lifted; and, through export-first policies, the internal economy has been more tightly integrated into the capitalist world market dominated by the North.

The Social Consequences of Adjustment

The debt crisis and structural adjustment contributed to widening the gap between the North and the South during the 1980s, with average income in the North reaching $12,510, ten times the average in the South.[6] All regions of the Third-World were squeezed by the debt crisis and structural adjustment, but the regions most ravaged were Latin America and Africa.

In Latin America, the force of adjustment programs struck with special fury, reversing the momentum of growth in the 1960s and 1970s. In a decade of zero growth, income inequalities among the worst in the world also worsened.[7] Peru, meanwhile, exhibits one of the worst cases of income distribution in the South, with 40 percent of the population getting 13 percent of the national income.[8] Tuberculosis and cholera, diseases that had been thought banished by modern medicine, came back with a vengeance throughout the continent.[9]

Among the more tragic consequences of widespread economic distress was the wearing down of the social fabric. Rising rates of violence in Colombia, for instance, cannot be divorced from its high debt-to-GNP ratio (38 percent).[10] Similarly, the rise of the Shining Path in both rural and urban Peru is inexorably tied to the travails wrought by structural adjustment in a country where the foreign debt comes to 50 percent of the GNP.[11] Africa has been even more devastated than Latin America. Total debt for sub-Saharan Africa comes to 82 percent of the GNP, compared to 36 percent for all developing countries.[12] Cut off from capital flows, battered by plunging commodity prices, wracked by famine and civil war, and squeezed by structural adjustment programs, Africa's per capita income declined by 2.2 percent per annum in the 1980s, and by 1990 it had reached its level at the time of independence in the early 1960s. If these bleak trends continue, the continent's share of the world's poor, now 30 percent, is expected to rise to 40 percent by the year 2000.[13]

With health systems throughout the continent "collapsing from lack of medicines,"[14] Africa is virtually defenseless against the AIDS epidemic, which now threatens to decimate the most productive strata of the population, those age twenty to forty-five. Yet, badly required resources needed to combat AIDS were or are being earmarked for debt servicing.[15]

Confronted with the social devastation created by their policies, technocrats at the World Bank and the IMF justify it as the bitter medicine that Southern countries have to swallow to gain economic health. But, after more than a decade of structural adjustment, the technocrats still lack an unqualified success story. They have no showcase in Africa.

Structural Adjustment and the Environment

Impoverishment has been among the main contributors to environmental damage in the Third World. This relationship is explained by a World Resources Institute (WRI) publication:

> Without jobs and without productive land, poor people are forced onto marginal lands in search of subsistence food production and fuelwood, or they move to the cities. Those who stay on the land are forced to graze livestock herds where vegetation is sparse or soils and shrubs are easily damaged and to create agricultural plots on arid or semiarid lands, on hillsides, in tropical forests, or in other ecologically sensitive areas.... As more and more people exploit open-access resources in a desperate struggle to provide for themselves and their families, they further degrade the environment.[16]

The toll on natural resources, continues the WRI account, takes many forms,

> including soil erosion, loss of soil fertility, desertification, deforestation, depleted game and fish stocks from overhunting and overfishing, loss of natural habitats and of species, depletion of groundwater resources, and pollution of rivers and other water bodies. The result is to reduce the carrying capacity and productivity of land and its biological resources. This degradation further exacerbates poverty and threatens not only the economic prospects of future generations but also the livelihoods, health, and well-being of current populations.[17]

The debt crisis and structural adjustment deepened the correlation between poverty and environmental degradation, as poor countries with already fragile ecosystems were forced to accelerate exploitation of valuable natural resources to earn foreign exchange to repay their debts. For instance, most of the top fifteen Third-World debtors have tripled the rate of exploitation of their forests since the late 1970s, a phenomenon undoubtedly related to the pressing need to gain foreign exchange to make interest payments. Indonesia and Brazil, two heavily indebted countries that also happen to contain much of the planet's remaining forests, saw their rates of deforestation increase by 82 percent and 245 percent respectively.[18]

Two countries, Ghana and the Philippines, illustrate the damaging environmental consequences of faithfully implementing structural adjustment programs. Ghana's structural adjustment program emphasizes intensified export production of cocoa to gain foreign exchange. But the rise in cocoa production was accompanied by a 48 percent decline in the world cocoa price between 1986 and 1989. Burdened with deteriorating terms of trade, Ghana was forced to even greater indebtedness to cover its burgeoning trade deficit, with its external debt rising from $1.1 billion in 1988 to $3.4 billion in 1988.[19]

To make up for declining foreign-exchange earnings from cocoa, the government moved to revive commercial forestry, with World Bank support. Timber production rose from 147,000 cubic meters to 413,300 cubic meters in the period 1984–1987, accelerating the destruction of Ghana's already much-reduced forest cover. At the current rate of deforestation, notes Fantu Cheru, Ghana could well be stripped of trees by the year 2000.[20]

Like Ghana, the Philippines has been one of the most faithful implementers of structural adjustment formula, being on this cure almost continuously since 1979 and paying as much as 25 to 30 percent of its foreign exchange earnings to servicing its debt. As in Ghana, the World Bank–IMF formula emphasizes gaining foreign exchange by stressing production for export, a policy that survived the transition from the Marcos dictatorship to the Aquino administration in the mid–1980s. Out of the almost $50 billion worth of products exported by the country between 1981 and 1989, traditionally resource-based exports accounted for almost $23 billion, or over 45 percent.[21]

Export-oriented growth, in short, was to a great extent a process of stripping a country once truly blessed by nature of its assets. By the end of the decade, the portion of the country covered by forest had declined from 50 percent in the 1950's to less than 18 percent, with most of the wood being exported to Japan; its coastal fish resources, much of it also exported, were depleted; and of its original 500,000 hectares of mangroves, the coastal breeding grounds of fish, only 38,000 hectares remained, much of the rest having been converted into fish or prawn farms geared mainly to producing for foreign markets.

Prawn farming in the Philippines provides a good illustration of the devastating environmental impact of debt-driven, export-oriented production. Not only has the creation of prawn farms to service the Japanese market entailed the destruction of mangroves, it has also disrupted traditional agriculture, owing to its dependence on a mix of salt and fresh water. On the one hand, the inflow of salt water threatens to lower the productivity of adjacent rice fields. On the other hand, so great are the prawn farms' demand for fresh water that rice farmers complain that there is not enough water to grow rice. Indeed, in some areas where land conversion to prawn farms has been extensive, water supply has dropped precipitously, prompting local authorities to ration it.[22]

Eternal Debt

Despite structural adjustment, debt rescheduling, and other financial mechanisms ostensibly geared to ending the debt crisis, the Third World's debt increased from $700 million in 1982 to $1.5 trillion by 1992. In order to protect themselves against probable defaults in repayment, most of the major banks established loan-loss reserves against their exposure and wrote down their loans, selling them in the secondary market at less than their original value. By 1991, the banks would have been inconvenienced but not seriously harmed if the South had defaulted on its debt of $1.2 trillion.[23]

What then is the logic of forcing the South to carry its load of debt, despite the tremendous social stress it is causing throughout the Third World and despite the overwhelming evidence that forgiving the debt would not disrupt the international economic order? The answer lies in Lester Thurow's observation that "Latin America and Africa cannot grow if they must service international debts as large as those that now exist. Too many resources have to be taken out to pay interest on those debts; too few are left for reinvestment."[24]

The point of the Reagan-Bush Counterrevolution was precisely to make it difficult for the South to leave a state of underdevelopment and again pose a threat to the North.

TRADE WARFARE: THE TAMING OF THE NEWLY INDUSTRIALIZING COUNTRIES

It was not only the more independent Third-World states that felt the lash of the Reagan–Bush counterrevolution but also the so-called Newly Industrializing

Countries (NICs), which included three of the staunchest allies of the United States during the Cold War: Taiwan, Singapore, and South Korea. Instead of structural adjustment programs, aggressive trade policy was used to put the NICs in line.

In the period from 1950 to 1980, the NICs had resorted to the same strategies of protectionism, tight controls on foreign investment, and strong state leadership that other Third-World governments had employed to achieve economic development. The United States, however, chose to overlook these deviations from market principles because of the priority it gave to its political alliance with states on the front lines of the Cold War.

In contrast to other Third-World governments, however, the NICs pursued export-led industrialization, attaching their growth to the U.S. market, which was the world's largest market and, until the early 1980s, one of the most open in the world. But from being the locomotive of East Asian growth, the United States during the Reagan years became an aggressively protectionist power.[25]

Tolerated during the global struggle against communism, the NICs' trade surpluses with the United States became the target of a concerted attack with the end of the Cold War. While world attention focused on the United States' increasingly problematic trade relations with Japan, the Reagan administration was subjecting the NICs to a multipronged trade offensive:

- In January 1989, the United States revoked the tariff-free entry of selected NIC imports under the General System of Preferences (GSP), with the rationale that the NICs have graduated from being developing countries. But even more damaging than this measure were the quantitative restrictions on NIC exports that go by the euphemism Voluntary Export Restraints (VERs).
- To make the NICs' exports more expensive and thus less appealing to American consumers, the United States forced the appreciation of the new Taiwan dollar and the Korean *won*.
- Protectionist measures were coupled with an aggressive drive to abolish import restrictions and lower tariff barriers to U.S. goods in the NIC markets. Liberalization and structural adjustment were now demanded from Korea and Taiwan. Behind the demand for liberalization was the threat to subject the NICs to a provision of the U.S Trade Act that required the U.S. president to take retaliatory measures against those officially tagged as "unfair traders."
- The federal government and U.S. corporations have teamed up to throttle unauthorized technology transfer to the NICs. While the U.S. government has sought to place restrictive convenants on intellectual property rights in the General Agreement on Tariffs and Trade (GATT), U.S. high-tech companies have initiated technological warfare against Korean and Taiwanese clone makers.

Aggressive trade and technological protectionism has been extraordinarily effective in subordinating Taiwan and Korea. Seeing the treatment of Korea, Taiwan, and other major Asian exporters of manufactured goods to the United States, the Third

World could not but fail to notice the hypocrisy of the U.S. position of preaching free trade and liberalization to the South while practicing aggressive unilateral protectionism and moving to create trading blocs like the North American Free Trade Agreement.

THE NEW CONTAINMENT

Having dismantled many of the co-optative mechanisms of the New Deal state, the Reagan and Bush administrations could not deal with the social stresses unleashed by its programs except through largely punitive measures. Punitive containment was the U.S. response not only domestically but also internationally. The South as the target of a new containment strategy, rather than the former Soviet Union, began to crystallize in the late 1980s. The image of "barbarians at the gate" evoked a siege mentality across the political spectrum, uniting conservatives with people from more liberal political traditions.

In his book *Millenium*, Jacques Attali, the well-known French socialist who heads the European Bank for Reconstruction and Development, capitulates to the anti-South mood and writes off the billions of people in the South as "millennial losers." Africa is a "lost continent," while Latin America is sliding into "terminal poverty." With no future of their own, says Attali, the peoples of the South can only look forward to "migrating from place to place looking for a few drops of what we have in Los Angeles, Berlin, or Paris, which for them will be oases of hope, emerald cities of plenty and high-tech magic."[26] What worries Attali is that the poor of the South "will redefine hope in fundamentalist terms altogether outside modernity. This dynamic threatens true world war of a new type, of terrorism that can suddenly rip the vulnerable fabric of complex systems."[27]

This image of protracted war is shared by the U.S. Presidential Commission on Long-Term Integrated Strategy, which believes that the struggle with the Third World will take the form of "low-intensity conflict—a form of warfare in which 'the enemy' is more or less omnipresent and unlikely ever to surrender."[28] But another outcome is possible in the 1990s. It can only come about, however, if people emerge who are brave enough to go against the tribal current by accepting the North's responsibility for the devastation of the South and working to promote a common agenda among the peoples of the North and the South based on their being the common victims of the global economic counterrevolution and the new containment.

CONCLUSION

The collapse of the South and the decline of living standards in the United States are two aspects of the same global counterrevolution promoted by two successive Republican administrations in the 1980s. The aim of this counterrevolution was twofold: to resubordinate the South within a U.S.–dominated international economic system and to dismantle the political economy of the New Deal that had established a fragile modus vivendi between U.S. capital and labor.

The debt crisis served as the opening wedge for structural adjustment programs that sought to roll back the state from economic activity, to eliminate protectionist barriers to Northern imports, to lift restrictions on foreign investment, to destroy the barriers set up by labor to restrain capital, and to integrate the local economy more tightly into the world economy. Against the up-and-coming NICs, the choice weapon was trade warfare, which aimed at restricting their exports to the U.S. market and prying open their domestic markets to U.S. goods and investment.

In the United States, tax policy was used to redistribute income from the majority to the rich minority; Pentagon capitalism was strengthened, partly with funds diverted from social services; unions were systematically assaulted by the state-capital alliance; and labor was devastated by the export of U.S. jobs to low-wage areas in East Asia and Latin America or by automation.

Ironically, a government pledged to restoring America's strength ended up gutting it by deficit spending for defense, which was financed largely with credits from the United States' main competitor, Japan; eschewing strategic economic planning in the name of market principles and putting America's economic future in the hands of corporations that were mainly interested in short-term profitability; and allowing the corporations to squander the United States' key resource in the competitive wars, its human capital.

Having abandoned the liberal containment strategy in the Third-World and dismantled the co-optative mechanisms of the New Deal state at home, the United States was reduced to strategies of punitive containment both at home and abroad. In moving toward a new grand strategy of containing the South, the United States was joined by many of the countries of the North, which sought to create security barriers against waves of immigrants and other "threats" from the devastated Third World and the former Eastern Bloc.

The challenge for progressives in the 1990s is to articulate a counteragenda based on the reality that the peoples of the South and the North share the tragedy of being victims of the same counterrevolution, which serves the interest of a global minority.

NOTES

1. Jeffrey Garten, *A Cold Peace: America, Japan, Germany and the Struggle for Supremacy* (New York: Times Books, 1992), 204.
2. Figures for the three regions are from KPMG, *The Asia-Pacific Region: Economic and Business Prospects* (Amsterdam: KPMG, 1988), 4.
3. Patrick Lloyd Hatcher, *The Suicide of an Elite: American Internationalists and Vietnam* (Stanford: Stanford Univ. Press, 1990).
4. Altaf Gauhar, "Arab Petrodollars," *World Policy Journal*, 4 (Summer 1987): 458.
5. The net outflow of financial resources from Africa to the IMF and World Bank between 1984 and 1990 came to almost $5 billion, with about $800 million going to the World Bank. World Bank statistics themselves show that between 1987 and 1991 the net transfer

of financial resources from Latin America and the Caribbean to the Bank came to over $2 billion. See Bade Onimode, "Critique of Orthodox Structural Adjustment Programs (SAP's) and Summary of the African Alternative" (Paper delivered at the Conference on People's Economics, Penang, Malaysia, 1991) 27. See also World Bank, *Annual Report 1991* (Washington, D.C.: World Bank, 1991), 139.

6. United Nations Development Program (UNDP), *Human Development Report 1991* (New York: Oxford Univ. Press, 1991), 23.
7. Ibid., 34.
8. Ibid.
9. Robin Wright and Doyle McManus, *Flashpoints: Promise and Peril in a New World* (New York: Alfred Knopf, 1991).
10. UNDP, *Human Development Report 1991,* 18.
11. Ibid., 154.
12. Ibid., 155.
13. Ibid., 23.
14. United Nations, *Financing Africa's Recovery: Report and Recommendation of the Advisory Group on Financial Flows for Africa* (New York: United Nations, 1988), 17.
15. UNDP, *Human Development Report 1991,* 155.
16. World Resources Institute, *World Resources: A Guide to the Global Environment,* 1991–93 (New York: Oxford Univ. Press, 1992), 30.
17. Ibid.
17. 18. Susan George, *The Debt Boomerang* (Boulder, Colo.: Westview Press, 1992), 11.
19. Fantu Cheru, "Structural Adjustment, Primary Resource Trade, and Sustainable Development in Sub-Saharan Africa," *World Development,* 20 (1992): 507.
20. Ibid.
21. Freedom from Debt Coalition, *Debt and Environment in the Philippines* (Quezon City, Philippines: Freedom from Debt Coalition, 1992), 14.
22. Asian Development Bank, *Economic Policies for Sustainable Development* (Manila: Asian Development Bank, 1990), 49.
23. Robin Broad, John Cavanagh, and Walden Bello, "Development: The Market Is Not Enough," *Foreign Policy,* 81 (Winter 1990): 151.
24. Lester Thurow, *Head to Head: The Coming Struggle Among Japan, Europe, and the United States* (New York: William Morrow, 1992), 215.
25. Garten, *A Cold Peace,* 203.
26. Jacques Attali, *Millennium: Winners and Losers in the Coming World Order* (New York: Times Books, 1991), 74.
27. Ibid.
28. Presidential Commission on Long-Term Integrated Strategy, *Discriminate Deterrence* (Washington, D.C.: Government Printing Office, 1988), 15.

21

Trading Away the Environment

Free-Trade Agreements and Environmental Degradation

Mark Ritchie

TWO COMPETING VISIONS vie for support concerning the economic future of the planet. One approach, often referred to as sustainable development, calls for social, political, and economic initiatives to protect the environment. This approach emphasizes the preservation of our soil, air, water, wildlife, and biodiversity in order to ensure economic security.

Sustainable development emphasizes less chemical- and energy-intensive industrial, forestry, fishing, and farming practices, and marketing practices that place a high priority on reducing the time, distance, and resources used to move products between production and consumption. In addition, it supports minimizing processing, packaging, and transportation in order to reduce waste and other ecological problems created by consumption.[1]

A competing vision, often referred to as the free-market, free-trade, or deregulation approach, pursues economic efficiency based on two principles. The first is to pay as little as possible for raw materials, and for the labor used to transform these raw materials. The second is to charge as much as possible to consumers, through monopolies and restrictive business practices. This approach externalizes almost all social, environmental, and health costs, ultimately paid for by today's taxpayers and by future generations. Proponents argue against any government intervention in daily business activities, claiming that government regulation diminishes economic efficiency. The multinational corporations involved in the buying and selling of goods on a global basis heavily promote free-market and free-trade policies, as do the international banks who, primarily, have financed these developments.

Debate between these two conflicting views has become a central global argument, intensifying subsequent to the Earth Summit in Rio de Janeiro in June 1992, officially called the United Nations Conference on Trade and Development. Free-trade advocates, led by the multinational corporations, argue that the best way to protect the environment is to eliminate government environmental regulations and to let the free market protect the planet. This view, while broadly rejected by governments of both the North and the South, is the ideological and practical view of powerful global corporations. This debate is just beginning.

The modern free trade vs. the environment debate in the United States sharpened in the early 1980s. During those years, the Reagan administration implemented the most free-trade–oriented economic programs since the 1920s. A close examination of one area of economic policy, agriculture policy, provides a clear insight into the politics of this debate.

There were at least four major effects of these free-market policies. First, they forced a huge number of farmers out of business. While record numbers of farmers went into bankruptcy, food processors earned record profits.

Second, government costs soared. Low prices meant huge deficiency-payment costs. The broader rural economic crisis caused by farm bankruptcies forced the government to bail out thousands of rural businesses, banks, and ultimately the entire farm credit system.

Third, many farmers reacted to falling farm prices by further intensifying their production methods. They hoped to make up in higher volume for the lower prices, but the increased use of chemicals and fertilizers only added to environmental and public-health problems. This intensification created enormous surpluses, forcing the Reagan administration in 1983 to impose one of the largest, most expensive, and most environmentally damaging land set-aside programs in U.S. farm history.

Fourth, the total value of U.S. farm exports declined sharply. Some farm policy analysts had warned that inelasticity in world food markets would mean that demand for food would remain relatively constant despite sharp drops in prices. Lower farm prices, they argued, would reduce the total value of U.S. farm exports, especially in the grains sector. But agribusiness economists convinced Congress that lower prices would drive other exporting countries out of the world market. These economists promised huge growth in export volume, enough to offset losses due to low prices.

POLITICAL REACTION TO FREE-MARKET FARM POLICIES

Farmers and small-town residents reacted sharply to Reagan's free-market policies, blocking foreclosure auctions and occupying government offices and banks. Rural America demanded an end to the destruction of their farms, families, livelihoods, and communities.

Consumer and environmental groups protested the safety of food and the ecological impact of chemical and energy-intensive production methods encouraged by free-market policies. The National Toxics Campaign, for example, launched a nationwide effort to change federal farm policies in ways that would reduce the use of chemical and energy-intensive production methods.[2] They advocated farm programs that would set farm prices at levels equal to the full cost of production, including all the environmental costs, while limiting production to the amount needed to balance supply with demand.

Agrichemical companies feared that these new proposals could lead to ever stricter pesticide regulations, particularly laws that greatly increased companies' financial liability for harm to workers, farmers, and communities that occurred during the production, storage, or application of their products. To avoid these regulations and liabilities, many chemical companies and corporate operators moved the production of the most dangerous products overseas.

Reacting to this sharp increase in overseas production of U.S. food supplies, some states and the federal government imposed increasingly stricter pesticide-residue regulations on imported foods. By 1989, as much as 40 percent of imported food items inspected by the U.S. Food and Drug Administration were rejected for reasons of unsafe chemical residues, contaminate levels, or other violations of U.S. standards.[3]

The problems created by these free-market policies generated a rebellion in both the countryside and in the cities. Corporate agribusiness and the agrichemical companies who had benefited the most from the free-trade approach feared that a political backlash might result in its dismantling.

COUNTERING THE BACKLASH

Agribusiness explored ways to counter the backlash. Food companies and exporters wanted to ensure that farm prices would not be returned to cost-of-production levels, thus ensuring the easy importation of cheap food from abroad. Agrichemical companies wanted to block any new local, state, or federal pesticide regulations.

One of the most creative strategies designed by agribusiness to counter the backlash was the decision to move policy-making on these issues out of the hands of state legislatures and Congress and into the arena of international trade negotiations. Under special fast-track procedures, the White House can negotiate away almost any aspect of domestic law that has any impact on trade. Under the fast-track rule, the White House negotiates in secret and presents the final result to Congress. Congress is barred by law from changing any aspect of the agreement. They can merely rubber-stamp it yes or no.

Using the fast-track procedure, the White House can call almost any social or environmental regulation a trade barrier. For example, food-safety standards could be termed "trade barriers" and then dismantled under the guise of "liberalizing trade." New rules for international trade could even roll back pesticide and other environmental regulations, while prohibiting restrictions on imported foods.

There are three important recent trade negotiations in which this strategy of moving decision making out of the hand of Congress can be clearly examined:

Canada-United States Free Trade Agreement (CUSTA), the North American Free Trade Agreement (NAFTA), and the General Agreement on Tariffs and Trade (GATT).

The first of these agreements, the Canada-U.S. trade agreement, was completed in 1988. It forced Canada to weaken regulations on pesticides and food irradiation, while U.S. laws on asbestos were suddenly open to challenge. A key feature of the accord gave the United States long-term and preferential access to Canadian oil, gas, and uranium resources that will extend the life of the nuclear-power industry and further hook the United States on petroleum-based fuels and industrial raw materials. The result will most likely be an acceleration of our problems with acid rain, ozone depletion, and global warming.

Under special fast-track procedures, Congress did not have the right to amend or fully debate this treaty, which was negotiated in secret. Such an approach neglects the broader interests of the public and subverts constitutional authority.

EXTENDING CUSTA TO NAFTA AND BEYOND

Almost before the ink was dry on the U.S.-Canada agreement, CUSTA, the transnational corporations aggressively pursued their global deregulation strategy by extending the CUSTA Agreement to Mexico, creating a North American Free Trade Agreement (NAFTA). This is the next step in a plan for a single Western Hemispheric free-trade zone, the "Enterprise of the Americas Initiative" (EAI) announced by the Bush administration in 1990. The EAI offers debt-for-nature swaps to relieve the official debt of Latin American countries who agree to adopt free-trade policies including environmental deregulation, free trade, foreign investment by transnational corporations, structural adjustment, and major cuts in social spending.

The lessons we learn from CUSTA and NAFTA can also be applied to the GATT. The GATT talks, involving over one hundred nations, have been marked by the same problems, such as the lack of environmental- and economic-impact assessments, the need for greater democratic participation, and the need for policies that consciously promote sustainable development.

UNDERMINING DEMOCRACY

One of the most important public concerns is whether free-trade negotiations will undermine democratic institutions in the United States and national sovereignty worldwide. Understanding how NAFTA and GATT undermine sustainable development requires examining how they undermine democratic institutions in the United States and enable global corporations to strengthen their control of the world economy by rewriting the rules of international trade. A new form of colonialism, these free-trade negotiations permit not only a continuing shift of wealth from the poor to the rich but also weaken each country's opportunities to protect its citizens and the planet from negative environmental consequences.

IMPACT ON LOCAL DEMOCRATIC CONTROL AND INITIATIVE

Over the past decade, the only real improvements in environmental protection have come "from the bottom up." Local governments have taken the initiative to strengthen consumer- and environmental-protection laws and regulations, often in response to organized pressure from grass-roots groups. These local efforts spread to other towns and regions, eventually forcing the state or provincial and federal governments to adopt similar measures. For example, southern California has led the way on clean-air regulations, which were then adopted by New York and other New England states, before being introduced in Congress as national standards.

A key political objective of all current trade negotiations, including the CUSTA as well as GATT, is to make it difficult, if not impossible, for subnational units of government to impose environmental and consumer safety standards that are more stringent than an international norm on any items that flow in world trade. Under current free-trade proposals, stricter state and local laws can be called "non-tariff barriers to trade" subject to challenge under the provisions of the agreements.

Although the long-term impact of free trade on local and state initiatives and powers in the environmental arena remains unclear, these provisions are placing actual limitations on states' rights and are creating an equally chilling response by states to the increased competition for jobs and investments. The ability of companies to move to a new location to avoid environmental regulation is discouraging many local and state governments from moving forward with new regulations to protect the environment. Citizens have a more difficult time advocating for change when discussion is removed from traditional democratic institutions, already favoring business interests, to a more exclusive and secretive forum.

The ecological crisis has created a danger so life-threatening that citizens around the world are demanding a thorough reorientation of our economic, political, and military structures and their right to participate in the global policy-making process to ensure that these changes are made. A revolution in telecommunications technology makes their participation both inexpensive and efficient. The only major barrier to democratized global policy-making is political opposition from those who benefit from the current authoritarian approach. Perhaps the energy created by the current environmental crisis can break down these political barriers and open up the entire global policy-making system—economic, ecological, and military—for effective citizen participation.

Ordinary citizens possess a great deal of knowledge about real situations and real options for solutions; they will be asked to make great sacrifices in both their lives and their standards of living in order to reverse the crisis. Unless the democratic process enables citizens to play a role in establishing necessary global policies, they will not accept or embrace the required changes. Unless the global process allows citizens some control of the decisions that create major changes in their lives, they will not willingly comply. The only options may be to use force to make citizens comply, or to abandon all political efforts to tackle these problems in hopes that the invisible hand of the marketplace will find a solution. However, the planet may not survive long enough to see if either the military or unbridled corporations are better suited to this challenge than political democracy.

The limits on government intervention heighten threats to sustainable development policies because NAFTA excludes social and environmental issues from negotiations. When the goal is economic efficiency, human, labor, and environmental rights disappear from the decision-making agenda.

NAFTA THREATS TO SUSTAINABLE AGRICULTURE

There are two main threats to sustainable agriculture in the NAFTA negotiations. The first is the stated objective of increasing the scale of production. Specific provisions in the text will lead to both increased corporate concentration in the processing sector and the further expansion of large-scale "factory farms" in the United States, Canada, and Mexico.[4] With no restrictions on resource extraction, Latin American governments will find it difficult to control industrial waste and the use of pesticides.

Another stated goal of free-trade agreements is the elimination of each government's ability to regulate the importing and exporting of goods. If, as a result of NAFTA, state and national governments cannot regulate the flow of goods across borders, farmers, consumers, workers, and the environment will suffer.

MAKING ORGANIC FARMING UNSUSTAINABLE

Free trade between the United States and Mexico may deliver a double whammy to organic farmers on both sides of the border. First, the general lowering of prices on commercially grown fruits and vegetables will make it hard to charge the prices needed to cover organic growers' additional costs. Second, expansion of fruit and vegetable production in Mexico will increase the overall use of chemicals, further disrupting and interfering with natural pest-control patterns. Organic farmers cannot apply pesticides to control pests driven to their fields by their neighbors' spray. Since organic farmers are dependent on natural predators for their own biological pest management, any increase in chemical spraying on neighboring farms will negatively affect their efforts to use biological pest management.

DESTROYING MEXICAN FAMILY FARMERS

One of the major demands of the multinational grain companies based in the United States is unlimited access for their exports of corn and other grains to Mexico. Almost three million Mexican peasants grow corn and sell this crop at price levels set high enough by the government to insure that they have enough cash income to survive. This system requires that the Mexican government carefully regulate imports to insure that this price level is not undermined.

Economists in both Mexico and the United States predict that, if the grain companies are successful in their efforts to force open the Mexican corn market, the price that will be paid to Mexican peasants will fall dramatically, forcing one million or more families off their land. Most of these families have worked at some time in the United States, so it is assumed that many will head north in search of either farm-worker jobs in the countryside or service-sector work in the major cities. Others will head to Mexico's urban areas, already dangerously overpopulated.

The United States, too, has used import regulations to sustain a domestic agricultural sector. For example, through the Meat Import Act of 1979, Congress established strict controls on the level of beef imports allowed into this country. But fast-food hamburger retailers have pushed the federal government hard to make sure that any NAFTA will abolish or weaken these controls, allowing them to import more hamburger meat. Since beef can be produced most cheaply on cleared rain-forest land in southern Mexico, a sharp increase in U.S. beef imports from this region would cause an acceleration in the destruction of the rain forest. A further worry is that Mexico will simply import its beef for domestic consumption from the rain-forest regions in Central and South America, freeing up its own beef production for shipping to the United States.

WEAKENING CONSUMER-HEALTH AND ENVIRONMENTAL-SAFETY STANDARDS

Few issues have caused as much conflict in the agricultural trade talks as the wide differences between each nation's food-safety and environmental standards. Corporations are lobbying for new GATT rules that could both limit the right of nations to set stricter standards and allow federal governments to pre-empt state pesticide and food safety legislation. This plan, referred to as "harmonization," limits the right of nations to impose consumer protection regulations on imported foods through the following procedures:

- Nations that attempt to set higher food-safety standards than those recognized by GATT would be subject to challenge by other countries on the grounds that it these higher standards would exclude their products from being imported. Countries with the higher standards would have to defend these standards or be subject to retaliation or have to pay compensation to the objecting countries.
- "Scientific evidence," as opposed to social or ethical criteria would be the only consideration in human-health and environmental regulations applied to imports. No social, economic, religious, or cultural concerns could be used to set import standards, no matter how important. A scientific court or adjudication panel, unaccountable to the public, would make decisions in relation to economic goals and objectives. Many nongovernmental organizations in the United States have

had bad experiences with faulty or dishonest science, ranging from the suppression of information about the dangers of silicon breast implants to promises of risk-free nuclear power. The prospect of seeing democracy undermined by a global science court is truly alarming.

- The Rome-based United Nations agency Codex Alimentarious, heavily influenced by executives from large chemical and food companies, has been chosen to set standards adopted by GATT. In all likelihood, these standards will be lower than most countries'.

If harmonization is prescribed by GATT, then national-government attempts to enforce domestic standards stricter than those recommended by Codex on pesticide residues on imported food (or on other food-safety concerns) could result in GATT-sanctioned trade retaliation or in demands for compensation to exporting countries. For example, a food item imported into the United States is banned under current legislation if it is found to have DDT residues above extremely low "background" levels. However, since Codex has set Maximum Residue Levels (MRLs) for DDT many times higher than the those the United States currently permits, disputes may arise between nations exporting foods with Codex-permitted DDT residues and the United States. The exporting nation could take this issue to a GATT dispute panel, who would compare United States limits to Codex's. The stricter U.S. standards could be ruled "GATT illegal." If the U.S. Congress subsequently refuses to revise the statute to meet GATT specifications, the United States would be subject to trade retaliation.

Many public-health advocates around the world fear that the GATT-enforced harmonization policy will be an instrument with which to overturn or weaken pesticide regulations and food-safety laws in the United States and elsewhere. Harmonization will restrict the ability of local and regional governments to set environmental- and consumer-protection standards. For example, the citizens of California have voted to prohibit the use of any carcinogenic pesticides on foods grown or sold in the state. But under harmonization this law could not be enforced on foods imported from overseas without the possibility of trade retaliation. Thus, generations of social and environmental advances could be eliminated in the United States and constrained in Mexico.

TOWARDS SUSTAINABLE TRADE

On the world scale and among the three nations of NAFTA, reform is needed in both commercial and political relations. The debates surrounding both GATT and NAFTA are unique opportunities to address these concerns. A positive outcome will require defeating global deregulation and the politics of the new world order.

Problems in the current trading regimes that need to be addressed include varying food-safety standards. A positive trade and development agreement would set minimum standards ("floors") for regulations rather than "ceilings." Any comprehensive development treaty must explicitly outlaw export "dumping," the exporting of goods by corporations at prices below the cost of production. U.S. and European grain-trading corporations regularly dump grain and dairy products at half the cost of production. This practice, which is destroying food self-sufficiency in poor countries and ruining family farmers everywhere, must be stopped.

Advocates of sustainable development everywhere in the world are exploring a wide range of trade-related policy issues. They can see the urgent need to ban food-product dumping in order to protect small farmers in both the North and the South. They recognize the need to ensure that the full costs of production, including environmental costs, are considered in the setting of prices. Otherwise, global food stocks will no longer be sufficient to handle the inevitable emergencies.

As a consensus evolves, we must accelerate our organizing. Agriculture groups from the United States must work with their colleagues from around the world toward the goal of establishing a common set of basic demands and solutions. This common agenda must then be promoted aggressively to all governments and to the public at large.

During the last decade there have been three breakthroughs in our understanding of the interrelationships between economy and ecology. The first is the inseparable connection between the environmental balance of the natural world and the modern industrial economy. Agriculture is at the center of this connection. Our relationship to the land—how we treat it, who lives on it, and who shares in its fruits—are central issues in our quest for a sustainable future. Close coordination between economic policy and environmental policy is a fundamental requirement for sustainability, both ecological and financial.

The second breakthrough is the acknowledgment that most ecological issues are global, respecting no boundaries. International cooperation and coordination in addressing ecological dangers is becoming an absolute necessity for human survival. Building a sustainable agriculture system in one state or region is insufficient and impossible. We need global agricultural policies that support, enhance, and enable the development of ecologically and economically sustainable agriculture in every region of the planet. This means that we must have both regional and global trade agreements that go beyond outdated theories of free trade to embrace the policies necessary for sustainable development, including labor, human-rights, environmental, debt, and economic-development issues.

Such dramatic shifts in policy, however, require a third breakthrough: that we change the underlying assumptions that shape our nation's trade policies, especially the idea that the earth and its natural resources can be used and abused endlessly. For several thousand years, a similar assumption was made about human beings: that they could be used and abused endlessly through slavery. At the end of the nineteenth century, after one hundred years of intense political organizing, slavery was finally outlawed in most countries. Now, we face a task much like that of these nineteenth-century abolitionists. We must lay to rest the idea that the earth's resources can be enslaved. There will be disruptive economic consequences, as there were with the abolition of slavery, but these disruptions cannot justify delay. The survival of future generations depends upon our success today at achieving sustainable life, balancing the economic and ecological relationships between people and the planet.

The controversy and debate created by the current trade negotiations must be translated into momentum for establishing new and more just relations among all nations. Nothing less can be accepted if we are serious about the survival of the planet.

NOTES

1. For additional views on sustainable development, see: Wes Jackson, Wendell Berry, and Coleman, eds. *Meeting the Expectations of the Land* (San Francisco: North Point Press, 1984); Benson and H. Yogtmann, eds., *Towards a Sustainable Agriculture* (Oberwill, Switzerland: Verlag Wirz AG, 1978); and Herman Daly, *Free Trade, Sustainable Development and Growth: Serious Contradictions* (Washington, D.C.: World Bank, 1992).
2. John O'Connor, *Shadow on the Land* (Cambridge,Mass.: National Toxics Campaign, 1988).
3. Subcommittee on Oversight and Investigations of the Committee on Energy and Commerce, U. S. House of Representatives, *Hard to Swallow: FDA Enforcement Program for Imported Food* (Washington, D.C.: Government Printing Office, 1989).
4. For additional details on the agricultural text of NAFTA, contact the Institute for Agriculture and Trade Policy, 1313 5th St., S.E., #303, Minneapolis, Minn. 55414.

22

Economics and Environmental Justice

Rethinking North-South Relations

Martin Khor Kok Peng

FOR SOME, THE cold war is over, the world can breathe easier, democracy is on the ascent, and the march of market forces will lead to great prosperity for all. For others, the emerging new world order is a cause for great depression: a return to an era of more direct colonialism; a negation of the principle of sovereignty over natural resources; a more stark willingness to reveal the primacy of power and self-interest over ethics and cooperation as the operating principle in world affairs; and the unembarrassed acceptance, as a way of life, of the coexistence of great wealth for a few and increasing misery for the majority of humanity.

On another level of debate, the world faces an ecological crisis of catastrophic proportions. This crisis challenges the notions of economic growth, development, international aid and cooperation, science and technology, knowledge systems and culture. For some, this crisis is a mere irritation, an externality or side effect of growth and technology. For others, this crisis calls for revamping production and technological systems and for the rapid deceleration of the wasteful consumption patterns of the world's elite. Some perceive the depletion of resources and the environmental crisis as an opportunity to mobilize the world community to new and higher forms of cooperation. For others, the increasing scarcity of resources is the basis for an intensified scramble for control over strategic raw materials, a scramble that will continue to lead to terrible conflicts in the next few decades.

From the perspective of groups working at the community level in developing countries, solutions to environmental problems at local, national, and international levels will fail, without linking ecological issues simultaneously with social issues of equity and economic issues concerning sufficient income and financial resources.

Solutions at the local level are intimately linked with national and international policies and structures. This is why many grass-roots groups in the Third World allocate some of their resources to international networking.

Increasingly severe economic crisis in most parts of the Third World accompanies the global environmental crisis. Per capita incomes in most African and Latin American (and some Asian) countries have been falling since the 1980s, and in some regions have declined to levels of twenty or thirty years ago. Poverty has increased, and with it health problems—such as cholera epidemics in Latin America and Africa.

These two phenomena—the global environmental crisis and socio-economic decline in the South—interconnect and result from an inequitable world order, unsustainable systems of production and consumption in the North, and inappropriate development models in the South. The operating principle of competition among economic institutions for profit in order to survive as economic entities makes economic growth a necessity. This principle operates within social systems that have a very unequal distribution of resources and incomes, thus resulting in uneven distribution of the benefits of growth and development. A small elite receives much of the world's output and incomes, while a large part of humanity has insufficient means to satisfy their needs: About one-fifth of the world's population, living in the North, obtains four-fifths of the world's income.

Yet the high rate of growth has led to the rapid depletion and contamination of resources, pollution, the proliferation of toxics, and climate-change threats. The extraction of raw materials and the production of cash crops have led to the depletion and exhaustion of natural resources, including energy, mineral, and biological resources. Moreover, the importation of inappropriate Northern technologies has progressively destroyed the more ecological indigenous production systems in the South. For example, modern agribusiness's chemical-dependent monocrop system, developed in the North, was transferred to the South in the so-called Green Revolution, an ecological nightmare. The diversity of crops and seeds has been eliminated, increasing the threat of crop failure and pest attacks. Ecological effects include deterioration of soil, loss of water resources, and loss of biodiversity.

The socio-ecological crisis of our times is thus the accelerating exhaustion and pollution of earth's resources through inappropriate technology and production processes that produce ever-increasing volumes of goods and services, the majority of which are channeled to fill the luxury wants of an elite. As those resources become scarcer, the justifiable demands of future generations will not be met.

From this perspective, the environmental and economic crises result from the same fundamental sources: the inappropriate and wasteful economic model of the North, the unequal distribution of resources and income at global and national levels, and the inappropriate development models in the South. The global link between the North's model and the South's model is obvious: the South's model is only a subset or a subsidiary of the dominant Northern economic model. The North's model was transferred to the South during colonialism, when the pattern of exchange was established between Southern raw materials and Northern capital and consumer manufactured products and accelerated in the postcolonial period through multilateral institutions that advised on macroeconomic policy and facilitated the continuation of the North–South production and trade pattern.

The postcolonial development model promoted by the World Bank and adopted by most of the Third-World countries coerced the developing countries to expand their exports of commodities. This has led to higher volumes of production, oversupply, lower prices, and continuous decline in terms of trade, with disastrous economic effects on the poor. In environmental terms this meant an acceleration in the depletion of natural resources, such as oil, forests, and minerals; the import of inappropriate Northern technologies that replaced the more ecologically sound systems of agriculture, fishery, and animal husbandry that existed in the South; and the transfer to the South of polluting industries, unwanted, unsafe products, and toxic wastes. The environmental crisis is really a side-effect of international economic relations. This same economic and development model exacerbates social problems like poverty, social inequities, and unbalanced development, as well as the depletion and contamination of resources. The World Bank has thus come under severe criticism in the Third World for having disbursed so much of its funds towards ecologically destructive projects, encouraging the production of export crops while people went hungry and opening vast tracts of forests for plantations and logging that could only fail to bring adequate returns.

THE LINK BETWEEN THE NORTHERN ECONOMIC MODEL AND SOUTHERN DEVELOPMENT

Given the pattern of world distribution of economic and technological power, the North with 20 percent of the world's population uses 80 percent of the world's resources and has an average per-capita income fifteen times higher than the South. The primary depletion and contamination of resources takes place in the North.

Analysts often say that production and consumption patterns have to change, a change that is politically impossible because no politician who advocates such change or diversion from economic growth would be elected. How then can we expect the much poorer South to change its economies? If a Northern politician is afraid to advise the public to have fewer cars per family, and to use less oil per car, can a Southern government be expected to do any differently in order to make way for two structural adjustments, one forced by external debt and the other dictated by ecological imperatives?

SHARING THE ECONOMIC BURDEN OF ADJUSTMENT

Since we must reduce the depletion of resources and also spend more to lower the ecological costs of pollution, waste, and climate change, then inevitably the volume of output must decline. For instance, to save the forests we have to reduce logging and reduce wasteful use of wood, which will require sharing the burden of economically adjusting to an ecologically sound pattern of production and consumption. This is a central international issue: How can we share the burden of adjustment more equitably both among and within nations?

At the international level, adjustment can occur in at least two ways. The first is if the powerful countries continue to say, "I am strong, you are weak; I want your resources that are getting more scarce, give them to me—too bad if you don't agree." In this solution, the poor continue to die off without help, sovereignty over resources erodes, and colonial rule again dominates parts of the world.

The second way is for citizens to push the governments of the world to agree on cooperation for mutual survival of their peoples. The North would thus say, "We have a mutual problem. We belong together as part of humanity; the overriding principle is that we all survive together. I am strong but perhaps I was wrong. In the colonial past and now in this present system, we have exploited nature, and yet many of you are still as poor if not poorer. And many of us frankly do not need to consume so many resources. Maybe we could adjust this unequal relation and have real partnership to save nature and thus ourselves."

As a matter of social justice, the North's responsibility in this new partnership should be obvious. In the postcolonial period, military conquest, extraction of natural resources, and an enormous flow of resource from South to North continues. Moreover, North-controlled multilateral institutions provided inappropriate advice or imposed inappropriate policies, such as increased commodity production or structural adjustment, which has resulted in social and ecological problems. Decisions made in a few major Northern countries, with no participation from the South, often result in enormous losses for the South: For example, realignment of exchange rates and interest-rate increases caused many Southern countries' external debt stock to rise and external debt-servicing flows to jump. Finally, it is predominantly the overconsumption of resources and the pollution emissions in the North that have caused the global environmental crisis.

This does not mean the South is absolved of all blame. In many parts of the South, there is a combination of corruption, political patronage, financial mismanagement, and the adoption of inappropriate technologies and environmentally unsound policies. But even in these seemingly national problems, Northern-controlled institutions play a role. Much of the misallocation of resources in the South can be traced to the bad economic advice or unfair conditions given by multilateral financial agencies and bilateral aid agencies.

There are strong moral and historical reasons why the North should take measures to reverse the South-North transfer of resources and to provide not charity but improvements in the South's terms of trade, to put life back into commodity pacts, to relieve the financial burdens weighing down the South, and to provide genuine aid for ecologically sustainable programs.

Is the North prepared to contribute to burden sharing? So far the North continues to concentrate its power. In the global economy, the global corporations strengthen their position vis-à-vis nation-states by pushing Northern governments to support their interests in negotiations on free trade. And Northern governments continually seek to prevent Third-World countries from imposing import restrictions or other obligations. As this occurs, Southern governments are obliged to allow foreign companies to enter and operate in their countries with minimal conditions across all social, economic, and cultural sectors. Local enterprises are unable to withstand the onslaught of the global corporations.

The South, by far the weaker party, cannot offer anything, as it has been on the receiving end of an unbalanced world order. Is the South the using environment as a bargaining chip, leverage to obtain more aid from the North? No, the South is essentially defending the last vestiges of its sovereignty over resources, particularly as the West provides more loans and aid to Eastern Europe.

NATIONAL AND INTERNATIONAL DEMOCRACY

Our goals in the South are about survival, humanity, and dignity. And democracy. A great deal of energy has been spent in the South in broadening the democratic spaces of our societies, in removing the barriers to people's participation, and in helping social movements regain their right to land and other resources, in order to promote their rights to good health and adequate nutrition, to safety, to housing, and to a sustainable environment. All these changes are necessary for both social justice and a sound environmental and development policy.

At the same time, the fight for democracy must extend to the international arena. A starting point would be to expand the democratic spaces in the international institutions that shape world policy and, through them, the national policies of our countries. The world economic order is obviously unbalanced, and there can be no progress toward a new international economic order without the democratization of the international economic institutions. Without a more balanced world economic order, there is little hope for any genuine partnership on the environment.

These institutions, including the global corporations, the international banks, the World Bank, the International Monetary Fund, and the General Agreement on Trade and Tariffs (GATT) should be made much more accountable to the public. Their decision-making processes must be opened for public participation and scrutiny. Local communities in our countries must also have the opportunity to participate in the design of programs and the monitoring of those programs' effects. The public has this right because the public suffers the consequences, whether it be the Bhopal residents dying from chemical poisoning, or the more than one hundred thousand farmers dying from pesticide poisoning annually, or the hundreds of millions of people suffering the social and economic effects of structural adjustment policies imposed by the World Bank and the International Monetary Fund.

SOME PROPOSALS

What needs to be done? Who will take the responsibility, make the sacrifice, bear the burdens, and implement the schemes? The acute problems of resource depletion, pollution and contamination, environmental health, and climate problems require urgent solutions at the international level. Solutions will involve fundamental changes in economic development models, throwaway life-styles, distribution of resources and income, and international political relations.

The economic model of the North needs to be discarded and a reevaluation of the international economic order undertaken. Basic questions include: how to change structurally the Northern models of production, income distribution, and consumption; how to promote ecologically sound and socially just development models in the South; how to structurally adjust the world economic institutions to promote fairer terms of trade and reverse the South-North flow of financial resources; and how to achieve a fair distribution in sharing the burden of adjustment necessitated by ecological imperatives, both between and within countries.

Changing the North, as previously noted, is the most pressing task. The North has to reduce wasteful production and scale down wasteful consumption. What kinds of institutional arrangements can be established within and between Northern countries to make these changes possible, to make the big corporations accept change? Can the adjustment burden be equitably shared?

In June 1991, a Ministerial Conference on Environment and Development was held in Beijing by developing countries. The conference declaration stressed that the inequities in current international economic relations had undermined the developing countries' abilities to participate effectively in global environmental efforts, and thus it is "imperative to establish a new and equitable international economic order" conducive to sustainable development in developing countries. The statement also reaffirmed the Third World's sovereign right over its natural resources and stressed that the developing countries must take the lead in eliminating environmental damage. The ministers called on the developed countries to provide new and adequate funds to enable the developing countries to cover the extra costs of tackling environmental problems and of implementing their commitments in international environment-related agreements. They proposed a special Green Fund to aid the Third World, to be "managed on the basis of equitable representation from developing and developed countries." Environmental concerns, economic change, and social equity must proceed hand-in-hand.

In the case of the South, what changes are required in international economic institutions to promote an international environment that facilitates the transition to sustainable development at the national level? What arrangements can be made to review and revise the policies of technical United Nations' agencies, such as the Food and Agricultural Organization, and private agencies to ensure that their programs conform to just and sustainable development? How can the World Bank, the IMF and the GATT be democratized, with fairer opportunities for Southern governments, nongovernmental organizations, and social movements to participate in decision making, planning, evaluation, and revising of policies and programs? How can their processes be made more transparent and publicly accountable? Because of their past poor record, we should oppose investing more resources and power in these institutions unless and until they are democratized and have proven that they have the technical competence to handle development and environment. At the same time, the Southern countries could agree to national adjustments favorable to the global ecology, such as a halt to destroying tropical forests, the conservation of biodiversity, and minimizing the use or production of harmful substances. The North could facilitate this process by paying higher prices for Third-World commodities, forgiving debt, and providing aid. In short, there is a need for a new era of international cooperation. Freed from control by nation-states, global corporations are moving rapidly to run the world system. Trade and markets are not free in this system; Southern countries are particularly dependent. Multilateral economic institutions such as the World Bank and the IMF are either facilitating or not stopping this process by which the Third World becomes further trapped in economic dependence and misery.

Finally, there is a need to establish a new or more-comprehensive international trading institution under the auspices of the United Nations, an institution whose objective would be to promote a more balanced North-South trade relationship, where the need for trade is tempered by the South's need for stronger domestic economies and a stronger position in world trade and economy. It would be inappropriate to promote the expansion of the supply and demand of Third-World raw materials, because such a demand depletes natural resources. The basic issue in commodities is how to reduce the volume of production and exports to conserve resources while raising prices to reflect their social and ecological values and thus enable the Third-World exporting countries to retain their export earnings. The shortfall in volume can be made up for by price increases, thus creating North-South cooperation in the sharing of the economic burden of adjusting to ecological principles.

CONCLUSION

Millions of people are enduring modernization, a code word for losing their control over their resources and their rights of determining their own future. Real or true development, for the Southern countries, means recognizing land rights, putting a stop to depleting resources such as forests; and introducing clean water, health facilities, and better schools.

Environmental and economic issues have to be resolved simultaneously, within the context of North-South relations, and within the operating principles of ecological sustainability and social equity. A fairer North-South balance at the international level would make it far easier for nongovernmental organizations (NGO)s in the South to facilitate genuine participation in achieving socially just and environmentally sound forms of development. At the same time, the North has to change. The battle for that adjustment in the North will be as difficult as it is necessary.

23

Solidarity with the Third World

Building an International Environmental-Justice Movement

Chris Kiefer and Medea Benjamin

THE LAST FIVE hundred years of global history tell a sorry tale of the theft and destruction of land and resources, all too often by people of Northern lands against people of the Southern Hemisphere. Colonialism, slavery, the introduction of export agriculture, and the unequal control of financial and informational flows have resulted in the passage of livelihood-sustaining resources into the hands of elite cultures and classes.

Conversely, there has always been resistance to this plunder, not only by victims but by outside sympathizers as well. Father Bartolome de las Cases and other priests denounced the first brutal exterminations of Native Americans. The U.S. abolitionists fought bravely against the abomination of slavery. Thousands of charitable organizations from industrial countries have sponsored every conceivable form of welfare program to address the worst manifestations of poverty in Third-World nations.

Contemporary efforts to link environmental-justice organizations of the North with indigenous struggles in the South are heirs to this long tradition. Yet today's struggle differs in several important ways. Industrial production has reached such a level as to endanger the viability of planet earth for humans and many other species. Technological advances in air travel and telecommunications have made it easier for people of the First and Third Worlds to interact directly. The paternalistic charity-consciousness of previous decades is gradually giving way to "solidarity," an understanding that it is in the self-interest of all of us to end poverty and environmental injustice.

This chapter examines the emerging relationship between two important kinds of actors in the environmental-justice movement: the nongovernmental human-rights and conservation organizations in the North, for whom environmental justice is a central concern, and the local and indigenous peoples' groups with whom they cooperate in the South.

The term "nongovernmental organization" (NGO), in its broadest sense, refers to any noncommercial, private, voluntary group whose purpose includes the pursuit of public issues, such as health, human rights, education, or environmental protection. This definition includes everything from single-issue groups, such as the International Rivers Network, to voluntary organizations with a general interest in human welfare, such as churches and trade unions. We shall use the term "NGO" in a more limited sense, to refer to private voluntary groups whose main agenda is the global environment (e.g., the Earth Island Institute, the Third World Network, the World Wildlife Fund, Rainforest Action, or the Sierra Club) or global justice and equity (e.g., Cultural Survival, Oxfam, or Survival International).

The interactions and relationships among Northern NGOs and indigenous organizations in less-developed areas is changing rapidly, including the problems, the knowledge base, the strategies, and the organizations themselves. Furthermore, the progress of the environmental-justice movement in the next few years will depend heavily on the success of such relationships. Our two-part thesis is, first, that the local and indigenous people threatened by big development must become politically effective if they are to survive. They must learn not only how to use the available local, national, and international machinery for environmental protection; they must also actively shape public opinion and create new laws and treaties where the existing ones are inadequate. This requires building solidarity among themselves and with potential allies in the North.

Second, while the environmental movement grows, destructive development exported by the industrial countries escalates rapidly. The struggle for the future of Third-World resources has entered a phase of intense political conflict in which the collective voice of previously voiceless people is pitted against the enormous wealth and influence of the major corporate and governmental players in the world economy. The outcome of this contest will be strongly shaped by the ability of the Northern NGOs to hear, understand, join, and amplify the voices of indigenous people. No clear strategy has emerged to accomplish this. The great diversity of peoples and problems guarantees that there will continue to be a certain invent-as-you-go character to this international cooperation. We report the ongoing search for workable ideas.

The 1970s and 1980s saw the dramatic failure of multinational and bilateral efforts to develop the economies of the less-industrialized countries. Among the conspicuous results of this failure are the accelerating destruction of tropical environments; the uprooting, repression, and decimation of indigenous peoples; the spread of urban slums; and the widespread hunger, illness, and unrest in the Southern Hemisphere.

Simultaneously, the Northern industrial nations have witnessed the rapid growth of environmental science and, with it, a deepening public consciousness of ecological problems both at home and abroad. Organizations promoting the protection of the environment, international human rights, and sustainable development have grown rapidly. Until the late 1980s, however, there was relatively little cooperation among groups in the various fields. Environmentalists sought to influence multinational organizations like the World Bank and United Nations Environment Program (UNEP) or the governments of less-industrialized nations, maintaining only intermittent communication with local victims of environmental destruction. Likewise, human-rights organizations appealed to local, national, and international legal bodies for protection of mainly political rights, with only incidental support for protecting environmental, cultural, and economic rights.

In the 1980s this began to change, for two reasons. First, the dramatic failure of top-down development led to increasingly vigorous opposition from the poor themselves. A famous example is the "Hug the Trees" (Chipko) movement of northern India. In 1974, after several years of disastrous floods due to the disruptive effects of logging in their traditional environment, a group of women from the village of Reni in the Garhwal Himalaya took action. They went into the forest and put their arms around the trees that the loggers were about to cut, refusing to move. From this small beginning, the movement spread throughout the area, attracting international attention. Eventually, the power of the movement became too much even for the Indian government and the logging interests. By 1978, vast sections of forest land had been set aside from commercial use, a direct result of Chipko.[1]

Inspired by the success of early grass-roots movements like Chipko, by the 1980s many local and indigenous communities had become much more active in the struggle to control their own environments. These struggles have brought them in conflict with national and local governments, international development agencies, global corporations, and Northern Hemisphere nongovernmental organizations (NGOs) that ignored local voices. Most U.S. NGOs in the environmental sphere strongly believe that preservation projects will not work unless the people most affected by them understand and approve of them and are involved from conceptualization to implementation.

Second, researchers, policy analysts, NGOs and grass-roots movements have become increasingly aware of the linkages among problems of environmental destruction, oppression, and top-down development. Holistic views of economic change quickly spread, so that by 1987, for example, the widely read Brundtland Commission report, *Our Common Future*, gave the United Nations imprimatur to a holistic notion of sustainable development. Since 1985, the details of the interrelationships have formed the basis of dozens of international and local conferences, sponsored by NGOs and governmental groups in ecology, economic development, human rights, and disarmament. These culminated in the United Nations Conference on Environment and Development (UNCED) in Rio de Janeiro in June, 1992. Although the Earth Summit failed to ratify a satisfactory global strategy, the mere fact of its occurrence was a modest victory for those pushing for greater environmental awareness. As a result of these developments, many U.S. NGOs in fields related to environmental justice have begun to change the way they operate in less-industrialized countries.

THE EMERGING ISSUES

There are so many indigenous and local organizations whose rights are threatened by economic development schemes in so many ways that it is difficult to characterize the problems from their point of view. In an effort to bring some coherence to the movement, in anticipation of the 1993 Year of the Indigenous People, a meeting of indigenous groups and their partner NGOs was convened in March 1992, under the auspices of the United Nations Economic and Social Council.

The summary list of critical needs drawn up at that meeting includes the representation of indigenous people in existing local, national, and international government; recognition of their legal customs and respect for their self-government; attention to their knowledge and skills in resource management and development; education of the general public on indigenous customs, rights, and needs; and recognition of their health needs, as well as their traditional health beliefs and practices.[2] This agenda resonates with prevailing democratic ideals of justice and calls for a complete overhaul of the world's legal, financial, and intellectual institutions.

From the viewpoint of Northern NGOs, this vision is very distant. In order to maximize the value of their limited resources, they must focus on those few projects that seem to promise major results quickly. The best issues are those that create vocal concern among the affected local and indigenous people and easily inspire concern among Northern Hemisphere people with knowledge, money, and power. The latter criterion is probably unavoidable, since almost by definition NGOs depend on the voluntary efforts and donations of large, diverse constituencies or on grants from large foundations whose boards are sensitive to public opinion.

The issues that qualify best tend to have certain common features. Most projects either have obvious consequences for humanity as a whole, such as global deforestation and species loss, or address dramatic suffering (and possible rebellion) among significant numbers of local people, as in cases of displacement due to land seizure for agriculture or hydroelectric development. The public saliency of other projects concerns the destruction of natural beauty, such as the decimation of large animals (e.g., caribou and elephants), the pollution of pristine waters, or the industrialization of park lands. To be effective, projects must address issues where the destruction is well documented.

Accordingly, global cooperation is proceeding most rapidly in five areas: protection of forests, protection of Arctic habitat, protection of traditional peasant land rights, protection of human communities from industrial pollution, and network-building among organizations of all kinds that deal with these issues.

Forest protection is by far the most developed issue area of North–South cooperation for environmental justice for several reasons. Environmental NGOs perceive the forest issue as one of the most critical because of the speed at which forests are being lost and the large-scale effects. Moreover, forest protection commands public attention because it appeals simultaneously to numerous concerns—global warming, soil conservation, biodiversity, aesthetics, and social justice for indigenous peoples. Forests also command the attention of a wide spectrum of activists because of the close links between the environmental aspects and human-rights aspects of forest preservation. Destruction of the forest necessarily means destruction of the indigenous culture. Finally, coalition-building of the sort we describe often works well when forest preservation is the issue. Forest people, being less dependent on industrial society than other local groups, have proven more resistant to co-optation and suppression; often they are good allies.

All these issues came to the fore in Honduras in 1991, when the Stone Container Company, a U.S. manufacturer, tried to exploit the largest contiguous forest in Central America. An unusually broad coalition was formed to stop Stone Container, including environmentalists, indigenous peoples, university students, trade unions, women's organizations, campesino groups, and human-rights activists. Such unity to protect the environment was unprecedented in Honduran history. The Hondurans then sought help from environmentalists, Central America solidarity groups, and human rights advocates in the United States. The pressure on both the Honduran government and Stone Container led to the withdrawal of the project.

The protection of Arctic habitat parallels the tropical forest issue in some important ways. The two zones are beautiful, vast, sparsely populated by humankind, and generally offer a stark contrast to the industrialized world. In 1991, several U.S. indigenous groups and NGOs joined with the Gwich'in tribal organization to protect the caribou herds of northern Alaska. The calving grounds of the Porcupine caribou lie within a 1.5-million-acre site called Tract 1002, adjacent to the 18-million-acre Arctic National Wildlife Refuge. The U.S. Fish and Wildlife Service refers to Tract 1002 as the center of wildlife activity for the entire refuge; nevertheless, in 1987 the U.S. Department of the Interior acceded to lobbying efforts by oil interests and announced the opening of Tract 1002 for exploration.

Since the Porcupine caribou have been the economic mainstay of the Gwich'in for thousands of years, the tribal leadership was watching for such a move. They called the first tribal council in one hundred years, formed the Gwich'in Steering Committee, and organized a protest campaign. The successful project forestalled petroleum development in Tract 1002.[3] In addition to research, lobbying, and coalition-building, the campaign worked with mainstream white groups to pressure the U.S. Congress.

A different type of environmental-justice issue is the protection of traditional cultivators from land expropriations and industrial development projects that threaten to forcibly displace them. Indigenous cultivators and herders are threatened in many areas of Central and South America, Africa, Indonesia, and India. The industrial world gives them little attention, despite the large number of people affected. One dam in Indonesia, Kendungombo, flooded nearly six thousand hectares of fertile land and displaced thirty thousand people; the Sardar Sarovar project, calling for thirty dams on India's Narmada River, threatens some two hundred thousand people.

The Sardar Sarovar dams project illustrates the level of international networking that has been achieved by environment and social-justice groups. In addition to Indian groups, the campaign to stop Sardar Sarovar includes groups from the United States, Germany, Australia, Holland, Finland, Japan, England, and Sweden. If the World Bank does not completely withdraw from this project, the coalition plans to launch a worldwide appeal to taxpayers, donor governments, and environment and social organizations to oppose all funding to the branch of the Bank that provides loans for such projects.

A WIDE RANGE OF STRATEGIES

Strategies for addressing these environmental and social-justice issues vary widely. One major role environmental groups have played is protecting the lives of local activists threatened because of their organizing work. After the murder of Brazilian labor and environmental activist Chico Mendes, the Rainforest Action Network and other environmental groups around the world protested his murder and became defenders of other activists threatened for their efforts to protect the rain forest and its inhabitants.

Some groups focus on gathering information about actual or planned development projects, making the data available to local and indigenous groups and to development organizations. They often lobby governments to stop destructive projects or policies. Environmental groups are pressuring the governments of Malaysia, Indonesia, Brazil, and other tropical counties to halt the logging and clearing of forest lands.

Another major target of lobbying is the World Bank. Many activists believe that their lobbying efforts have brought some policy changes but that resistance remains to their implementation. For example, "The World Bank's forest policy is changing; they require a full Environmental Impact Assessment on any project that might affect native peoples," says Kathy Fogel of the Sierra Club. "They have language on consulting with native peoples and with local environmental organizations. However, the Bank continues to support logging, directly and indirectly, in primary forests of all types."[4]

The Bank has withdrawn support from some projects, largely as a result of pressure from environmental-justice advocates. However, even successful lobbying efforts are often dashed when other banks or private lenders fund a project after the World Bank declines. Having successfully lobbied the Bank to withdraw support for a geothermal project, the Philippine Development Forum soon found the developer negotiating with a Japanese bank.

Another strategy has been organizing boycotts, such as the boycott of tropical timber products. Logging accounts for some 20 percent of the destruction of tropical rain forests around the world, and most of the timber is exported to industrialized countries. Current campaigns include banning the use of tropical timber in Europe, the United States, and Japan. In the United States, activists have been working to ban the use of tropical timber by cities, counties or states. By 1992, the states of New York and Arizona, as well as the cities of San Francisco; Santa Monica, California; Bellingham, Washington; and Baltimore had banned the use of tropical timber in public projects. Similar state legislation has been debated in California, Maryland, and Massachusetts, as well as in dozens of cities. Congress introduced legislation to require labeling on all imported tropical woods and wood products, making it much easier for consumers to avoid purchasing such wood.

A new tactic with enormous potential is related to boycotts: developing alternative trade in goods that bring income to native peoples and are environmentally sound, the leader of which has been Cultural Survival. "We realized that advocacy and publicity alone wouldn't save people," says Jason Clay, director of Cultural Survival Enterprises (CSE), "that most of our efforts and money should go to projects that indigenous people themselves design and run."[5] In 1989, Cultural Survival Enterprises was formed to help rain forest communities produce and market their products. Concentrating on trade in nuts, fruits, oil, resins, essences, pigments, and flours, by 1992 CSE was selling some $2.5 million in forest products to twenty-six companies making nearly forty products.

Other environmental groups have criticized Cultural Survival's project on the grounds that it ties indigenous people into the market economy and makes them dependent on foreign markets. "Binding the economic future of tribal peoples to the creation of ephemeral, foreign markets in nonessential luxuries such as ice-cream or shampoo with added rainforest ingredients will not solve their problems," says Stephen Corry, director of the London-based Survival International.[6] Cultural Survival counters that forest resources will be exploited one way or another, but that nontimber-marketing projects allow for indigenous control and sustainable management. "Without income from sustainably harvested forest projects," insists Marc Miller of Cultural Survival, "most forest residents would be forced to degrade their environment to meet their material needs—or abandon it to others who would [degrade it]."[7]

A means of preserving the environment that has been sharply debated is debt-for-nature swaps, whereby North American or European environmental organizations purchase part of a country's foreign debt in exchange for an agreement from the local government to create parks or bioreserves. The first such swaps were harshly criticized by indigenous groups for leaving them out of the process. As one group of indigenous peoples' organizations from the Amazon warned: "We are concerned about the 'Debt for Nature Swaps' which put your organizations in a position of negotiating with our governments for the future of our homelands. We know of specific examples of such swaps which have shown the most brazen disregard for the rights of indigenous inhabitants. We propose you swap 'Debt for Indigenous Stewardship' which would allow your organizations to help return areas of the Amazonian rainforest to our care and control."[8]

Some environmentalists also point out that the huge debts of less-developed countries overwhelm the modest sums involved in such swaps. According to a U.S. Congressional report, a total of twenty-six such swaps were carried out in thirteen less-developed countries between 1986 and 1990. The amount of debt saved was $126 million, which equals less than five hundredths of one percent of those countries' total debt. Two-thirds of the exchanged debt was in one country, Costa Rica, whose total debt was reduced by only two percent. Such exchanges do little to reduce the need of Third-World governments for hard foreign currency, or their incentive to sell their citizens' environmental heritage in order to obtain that currency.

While none of these strategies has proven to be a panacea for environmental injustice, as early experiments they are producing a useful fund of knowledge for future work, even when they fail to achieve human rights and a healthy environment. Indigenous groups and Northern NGOs learn from their failures as well as their successes, gaining a clearer sense of their own limitations and strengths, their own illusions and insights.

BUILDING EFFECTIVE NORTH-SOUTH COMMUNICATION

Achieving economic and environmental justice requires close collaboration with local indigenous and environmental groups. But it is often difficult for U.S. groups to figure out with whom to work. While in the early 1980s cooperation was hampered because most Northern NGOs did not know what local groups existed in any given location, the problem now often concerns sorting out the many known groups potentially affected by a given project. But as is the case the world over, organizations have disagreements over everything from strategies to finances to personalities. Groups may split into different factions, align themselves with different political parties, get co-opted by local elites or government agencies, or become dependent on the skill and commitment of one individual. Indigenous groups may have long-standing land disputes with each other, making it difficult to distinguish the "good guys" from the "bad guys." U.S. activists with personal experience working with Third-World groups tell us of the slow, often painful process of learning the "lay of the land" and avoiding entanglement in internal disputes. Other difficulties include barriers such as differences in language, culture, education, class, and access to resources. Furthermore, many Third-World groups hold a deep-seated suspicion of outsiders because of the Third World's history of colonialism. Oftentimes these groups' only contact with foreigners has been through global corporations or missionaries.

When U.S. environmental groups first tried to establish a relationship with groups in Honduras, for example, they encountered enormous mistrust. "We knew about the American business people who were here to steal our resources," recalled Gladys Lanza of the Coordinating Committee for Popular Organizations. "We knew about the American missionaries who were here to convert our souls. Our only contact with Americans had been negative. So of course we were suspicious."[9] Building trust can be a long, difficult process.

Third-World groups often believe that once they start working with U.S. groups—groups that have so many more resources and such greater access to power than they do—the Americans start to dominate the agenda. This was the case in Ecuador when U.S. environmental groups were supporting the efforts of Ecuadoran groups to stop the Conoco oil company from drilling in the Yasuni National Park and the Huaorani indigenous reserve. The Ecuadoran Amazon Campaign, a coalition of Ecuadoran environmental and human-rights groups, accused two U.S. NGOs—the Natural Resources Defense Council and Cultural Survival—of misrepresenting their views in negotiations with Conoco and of intentionally deceiving them in a "blatant display of ecological imperialism."[10]

"In general, we welcome the opportunity to collaborate with our North American colleagues," the Campaign members wrote in an open letter to the U.S. environmental community. "We greatly appreciate and fully support their efforts to lobby the U.S. Congress, the multi-lateral development banks and the multi-nationals on behalf of the Amazon Biosphere and its inhabitants. But we are confronted with two North American organizations who have purposely misled Ecuadorians and who insist on negotiating with an oil company on our behalf, without our consent, and without respecting our right to articulate our own needs and aspirations."[11]

Such problems indicate the extreme sensitivity that U.S. groups must have when dealing with Third-World organizations. Some activists are skeptical of the wisdom of Northern environmentalists trying to mount cooperative projects with local groups, given the difficulties of understanding local cultural and political subtleties. "[Working with indigenous groups] is often suspect and always a tricky process for Northern NGOs," says Lafcadio Cortesi of Greenpeace. He advocates a translator role, whereby Northern NGOs summarize and clarify the enormously technical reports and proposals used by official agencies and present the facts to indigenous people and local governments in a more usable form. One such form is the Tropical Forest Action Plan (TFAP). Local groups can then draw up plans based on their own needs and perceptions, and the NGOs can help evaluate the practical outcomes.

There is a fine line between supporting the agenda as defined by indigenous organizations themselves and *creating* the agenda. U.S. groups must constantly ask themselves whether they are crossing that line. This kind of sensitivity reduces the likelihood of local people being co-opted, for instance, by money, but it is also very difficult in practice. Because of their habitual task-orientedness, many Northern activists have difficulty seeing the point of discussions that are not already focused on a concrete project.

Many groups in the South also believe that Northern groups are more concerned about saving tropical rain forests than saving their own forests, or more concerned about overpopulation in the South than overconsumption in the North. If Northern NGOs really want to build effective working relationships with Third-World groups, they must develop a sensitivity to related problems in their own backyard. This will help them not only address the worries of Third-World groups, but will also help them develop a more holistic strategy for resource conservation.

Solid communication with Third-World groups is also hampered because the majority of the staff and constituency of most U.S. environmental groups is white and middle- or upper-class. This makes it more difficult for U.S. groups to identify with poor people of color—be they in the Mississippi Delta or the Brazilian Amazon—and vice versa. The recent push to get more people of color on the staff and boards of U.S. environmental groups, and the explosion of interest in the relationship between race, poverty, and environmental issues are positive signs that will indirectly have a beneficial effect on North–South communication.

An overall view of North–South communication on environmental justice reveals the great importance of attitudes and assumptions conditioned by culture and experience. Just as the negative experiences of many local peoples tend to burden such communication with mistrust, Northerners carry many unquestioned assumptions that have to be rethought as well. Americans, especially those active in humanitarian organizations, tend to overestimate the similarities of value and viewpoint among peoples—i.e., to see others in their own image. As products of a technocratic culture, they usually overvalue efficiency and technical knowledge at the expense of tradition and untutored experience. Similarly, many middle-class Americans tend to pay close attention to the *ideas* in a discussion, but overlook the more subtle but equally important emotional and social cues.

STRENGTHENING THE LINKS

Nevertheless, cooperation is growing rapidly, and organizations on both sides are accumulating valuable experience. Much of the knowledge that results from different campaigns, successful or not, is being passed on from region to region and refining the movement's strategies.

Another hopeful factor is that indigenous people themselves, and to some extent their partner NGOs in the North, have a great deal more knowledge of the local land, the local culture, and the process of their interaction than the bureaucrats and technicians who want access to their local resources. With greater coordination and international recognition, local people can have more ability to influence power.

The work thus far has also proven that governments, corporations, and international agencies are somewhat sensitive to the opinion of their constituents, stockholders, and customers. As the North–South movement grows, it will create a broad, well-informed and vocal public, alert to the connections between their own beliefs and interests on one hand and the violation of indigenous environmental rights on the other hand. The survival of the environment–social-justice movement will depend largely on the success of this endeavor.

NOTES

1. See Vandana Shiva and J. Bandyopdhyay, "The Evolution, Structure, and Impact of the Chipko Movement," *Mountain Research Development*, 6 (1986): 133-42.
2. United Nations Economic and Social Council, "Technical Meeting on the International Year for the World's Indigenous People," (a paper presented to the Technical Meeting on the International Year for the World's Indigenous People, 9-12 March 1992, Geneva, Switzerland).
3. Bedford, M. "Saving a Refuge," *Cultural Survival Quarterly* (Spring 1992): 38-42.
4. Telephone interview, 3 August, 1992.
5. Telephone interview, 5 August, 1992.
6. Knaus, H. "Debating Survival," *Multinational Monitor* (September 1992): 13.
7. Ibid., 14.
8. "Open Letter to Environmentalists, COICA," *Earth Island Journal* (Winter 1990): 2.
9. Telephone interview, 6 August, 1992.
10. Letter from "La Campana Amazonia Por La Vida," Quito, Ecuador, 25 May 1991.
11. Ibid.

Appendix

Principles of Environmental Justice

Adopted at the First National People of Color Environmental Leadership Summit, October 24–27, 1991, Washington, D.C.

Preamble

We, the people of color, gathered together at this multinational People of Color Environmental Leadership Summit, to begin to build a national and international movement of all peoples of color to fight the destruction and taking of our lands and communities, do hereby re-establish our spiritual interdependence to the sacredness of our Mother Earth; to respect and celebrate each of our cultures, languages and beliefs about the natural world and our roles in healing ourselves; to insure environmental justice; to promote economic alternatives which would contribute to the development of environmentally safe livelihoods; and, to secure our political, economic and cultural liberation that has been denied for over 500 years of colonization and oppression, resulting in the poisoning of our communities and land and the genocide of our peoples, do affirm and adopt these Principles of Environmental Justice:

1. *Environmental justice* affirms the sacredness of Mother Earth, ecological unity and the interdependence of all species, and the right to be free from ecological destruction.
2. *Environmental justice* demands that public policy be based on mutual respect and justice for all peoples, free from any form of discrimination or bias.
3. *Environmental justice* mandates the right to ethical, balanced and responsible uses of land and renewable resources in the interest of a sustainable planet for humans and other living things.

4. *Environmental justice* calls for universal protection from nuclear testing and the extraction, production and disposal of toxic/hazardous wastes and poisons that threaten the fundamental right to clean air, land, water, and food.
5. *Environmental justice* affirms the fundamental right to political, economic, cultural and environmental self-determination of all peoples.
6. *Environmental justice* demands the cessation of the production of all toxins, hazardous wastes, and radioactive materials, and that all past and current producers be held strictly accountable to the people for detoxification and the containment at the point of production.
7. *Environmental justice* demands the right to participate as equal partners at every level of decision-making including needs assessment, planning, implementation, enforcement and evaluation.
8. *Environmental justice* affirms the right of all workers to a safe and healthy work environment, without being forced to choose between an unsafe livelihood and unemployment. It also affirms the right of those who work at home to be free from environmental hazards.
9. *Environmental justice* protects the right of victims of environmental injustice to receive full compensation and reparations for damages as well as quality health care.
10. *Environmental justice* considers governmental acts of environmental injustice a violation of international law, the Universal Declaration On Human Rights, and the United Nations Convention on Genocide.
11. *Environmental justice* must recognize a special legal and natural relationship of Native Peoples to the U.S government through treaties, agreements, compacts, and covenants affirming sovereignty and self-determination.
12. *Environmental justice* affirms the need for an urban and rural ecological policies to clean up and rebuild our cities and rural areas in balance with nature, honoring the cultural integrity of all our communities, and providing fair access for all to the full range of resources.
13. *Environmental justice* calls for the strict enforcement of principles of informed consent, and a halt to the testing of experimental reproductive and medical procedures and vaccinations on people of color.
14. *Environmental justice* opposes the destructive operations of multi-national corporations.
15. *Environmental justice* opposes military occupation, repression and exploitation of lands, peoples and cultures, and other life forms.
16. *Environmental justice* calls for the education of present and future generations which emphasizes social and environmental issues, based on our experience and an appreciation of our diverse cultural perspectives.
17. *Environmental justice* requires that we, as individuals, make personal and consumer choices to consume as little of Mother Earth's resources and to produce as little waste as possible; and make the conscious decision to challenge and

reprioritize our lifestyles to insure the health of the natural world for present and future generations.

Selected Bibliography

Alston, Dana, ed. *We Speak for Ourselves: Social Justice, Race, and Environment.* Washington, D.C.: The Panos Institute, 1991.

Anthony, Carl. "Don't Abandon the Inner Cities." *Earth Island Journal* (Fall, 1991).

Bryant, Bunyon, and Paul Mohai, eds. *Proceedings of the Michigan Conference on Race and the Incidence of Environmental Hazards.* Ann Arbor, Mich.: Univ. of Michigan School of Natural Resources, 1990.

———. *Race and the Incidence of Environmental Hazards: A Time for Discourse.* Boulder, Colo.: Westview Press, 1992.

Bullard, Robert. D. ed. *Confronting Environmental Racism: Voices from the Grassroots.* Boston: South End Press, 1993.

———. Environmental just*ice and Communities of Color.* San Francisco: Sierra Club Books, 1993.

———. *People of Color Environmental Groups Directory.* Riverside Calif.: Univ. of California, 1992.

———. *Dumping in Dixie: Race, Class, and Environmental Quality.* Boulder, Colo.: Westview, 1990.

Bullard, Robert D., and Beverly Wright. "The Quest for Environmental Equity: Mobilizing the African-American Community for Social Change." *Society and Natural Resources* 3 (1990): 301–311.

Churchill, Ward. "Radioactive Colonization: A Hidden Holocaust in Native North America." (Unpublished article, 1992).

Durning, Alan. *Action at the Grassroots: Fighting Poverty and Economic Decline.* Paper no. 88. Washington, D.C.: Worldwatch Institute, 1989.

Environmental Careers Organization. *Beyond the Green: Redefining and Diversifying the Environmental Movement.* Boston: Environmental Careers Organization, 1992.

Faber, Daniel. *Environment under Fire: Imperialism and the Ecological Crisis in Central America.* New York: Monthly Review Press, 1993.

Gottlieb, Robert. *Forcing the Spring: The Transformation of the Environmental Movement.* Washington, D.C.: Island Press, 1993.

Krauss, Celene. "Community Struggles and the Shaping of Democratic Consciousness." *Sociological Forum* 4 (1989).

———. "Women and Toxic Waste Protests: Race, Class and Gender as Resources of Resistance." In *Environmental Justice and Communities of Color,* edited by Robert D. Bullard. San Francisco: Sierra Club Books, 1993 (forthcoming).

Mann, Eric, with the Watchdog Organizing Committee. *Los Angeles' Lethal Air: New Strategies for Policy, Organizing, and Action.* Los Angeles: Labor/Community Strategy Center, 1991.

McLaughlin, Andrew. "Ecology, Capitalism, and Socialism." *Socialism and Democracy* 10(Spring-Summer 1990): 71.

Mellor, Mary. *Breaking the Boundaries: Towards A Feminist, Green Socialism.* London: Virago, 1992.

Rensenbrink, John. *The Greens and the Politics of Transformation.* San Pedro, Calif.: R. E. Miles, 1993.

Ross, Andrew. *Strange Weather: Culture, Science, and Technology in the Age of Limits.* New York: Verso, 1992

Vandana Shiva, ed. *Close to Home.* Philadelphia: New Society Publishers, 1993.

———. *Staying Alive: Women, Ecology, and Development.* New Dehli: Kali Books, 1989.

Sontheimer, Sally, ed. *Women and the Environment: A Reader.* New York: Monthly Review Press, 1991.

Souza, Bonnie. "Justice or Ecocide: The Challenge Facing The Environmental Movement and Opportunities for Organizing in the Pacific Northwest." Masters Thesis, Univ. of Oregon, 1992).

Szasz, Andrew. *Environmental Protests and the Grassroots.* (Forthcoming).

Truax, Hawley. "Beyond White Environmentalism: Minorities and the Environment." *Environmental Action* 21 (1990): 19–30.

United Church of Christ Commission for Racial Justice. *Toxic Wastes and Race in the United States: A National Report on the Racial and Socio-Economic Characteristics of Communities with Hazardous Waste Sites* New York: United Church of Christ, 1987.

U.S. Environmental Protection Agency. *Environmental Equity: Reducing Risk for All Communities.* vol. 2. Washington, D.C.: Government Printing Office, 1992.

U.S. Environmental Protection Agency Office of Communications, Education, and Public Affairs. "Environmental Protection—Has It Been Fair?" *EPA Journal* 18 (March-April, 1992).

U.S. General Accounting Office. *Siting of Hazardous Waste Landfills and Their Correlation with Racial and Economic Status of Surrounding Communities.* Washington, D.C.: Government Printing Office, 1983.

Wallis, Victor. "Socialism, Ecology, and Democracy: Toward A Strategy of Conversion." In *Socialism: Crisis and Renewal,* edited by Chronis Polychroniou. New York: Praeger, 1992.

SELECTED PERIODICALS

Capitalism, Nature, Socialism: A Journal of Socialist Ecology
ATTN: Journals Department
Guilford Publications
72 Spring Street
New York, NY 10012

Environmental Action.
6930 Carroll Avenue
Takoma Park, MD 20912

Everyone's Backyard: The Journal of the Grassroots Movement for Environmental Justice
Citizen's Clearninghous for Hazardous Waste
P.O. Box 6806
Falls Church, VA 22040

New Solutions: A Journal of Environmental and Occupational Health and Safety
Work and Environment Program
Lowell University
Lowell, MA 01854

Race, Poverty, and the Environment: A Newsletter for Social and Environmental Justice
Earth Island Institute
Urban Habitat Program
300 Broadway
San Francisco, CA 94113

Contributors

WALDEN BELLO is the executive director of the Institute for Food and Development Policy in San Francisco. His books include *People and Power in the Pacific: The Struggle for the Post-Cold War Order* (1992), *Dragons in Distress: Asia's Miracles in Crisis* (1991), and *Development Debacle: The World Bank in the Philippines* (1982). He is also a fellow at the Transnational Institute in Amsterdam, and at the Center for Southeast Asian Studies at the University of California at Berkeley.

MEDEA BENJAMIN is executive director of the San Francisco-based Global Exchange and author or co-author of several books, including *Don't Be Afraid, Gringo: A Honduran Woman Speaks From the Heart* (1988), *No Free Lunch: Food and Revolution in Cuba Today* (1989), and *Bridging the Gap: A Handbook for Linking First and Third World Citizens* (1991).

ROBERT D. BULLARD is professor of sociology at the University of California at Riverside. He is the author of *Confronting Environmental Racism* (1993), *Dumping in Dixie: Race, Class, and Environmental Quality* (1990), and *People of Color Environmental Groups Directory 1992.* A leading expert on environmental justice, he is the author of numerous articles and monographs on environmental problems in communities of color as well as issues of housing, land use, industrial policy, industrial facility siting, and community development. Bullard was a planner of the First National People of Color Environmental Leadership Summit.

CESAR CHAVEZ was the Founder and President of the United Farm Workers of America. He died on April 22, 1993.

MARCIA A. COYLE is Washington bureau chief and Supreme Court correspondent for the *National Law Journal* in Washington, D.C.

BARBARA EPSTEIN is a professor on the History of Consciousness Board at the Univesity of California at Santa Cruz. She is the author of *Political Protest and Cultural Revolution: Nonviolent Direct Action in the 1970s and 1980s* (1991) and *The Politics of Domesticity: Women, Evangelism, and Temperance in Nineteenth Century America* (1981). Epstein is a regular contributor to *Socialist Review.*

DANIEL FABER is an assistant professor of sociology at Northeastern University, Boston, Mass. He is a member of the editorial collective of *Capitalism, Nature, Socialism: A Journal of Socialist Ecology* and has written numerous articles on environmental issues. From 1984 to 1990 he served as research director for the Environmental Project on Central America. His latest book is entitled *Environment Under Fire: The Ecological Crisis in Central America* (1992).

CYNTHIA HAMILTON is director of African and African-American Studies at the University of Rhode Island at Kingston. She is the author of numerous articles on environmental justice and a member of the Board of Organizers of the Labor/Community Strategy Center.

LOUIS HEAD is staff associate for the SouthWest Organizing Project in Albuquerque, a multiracial grass-roots organization whose mission is to empower the disenfranchised in the Southwest to realize racial and gender equality, and social and economic justice. He has been an activist and organizer for many years.

RICHARD HOFRICHTER is the executive director of the Center for Ecology and Social Justice and an associate fellow at the Institute for Policy Studies in Washington, D.C. He is the author of *Neighborhood Justice in Capitalist Society: The Expansion of the Informal State* (1987).

MARTIN KHOR KOK PENG is the director of the Third World Network and is the research director of the Consumers Association of Penang, Malaysia. He is also the director of the World Rainforest Movement and of the Asia Pacific Peoples Environmental Network. He is the author of several books, including *The Uruguay Round and Thirld World Sovereignty* (1990).

CHRIS KIEFER is an associate professor of anthropology at the University of California at San Francisco. He has written and lectured extensively on health, peace, and environmental issues.

YNESTRA KING teaches at the Eugene Lang College, New School for Social Research in New York. She is the author of *Eco-Feminism and the Reenchantment of Nature: Women, Ecology and Politics* (1991) and is a leading eco-feminist theorist.

CELENE KRAUSS is an assistant professor of sociology at Kean College of New Jersey and co-coordinator of the Women's Studies Collateral. She has written extensively about women and toxic wastes, with a focus on issues of race and class.

WINONA LADUKE is the co-chair of the Indigenous Women's Network, a grass-roots network of Native and Pacific island women, and director of the White Earth Land Recovery Project, a reservation-based land and cultural preservation organization in White Earth, Minn. Her extensive writings have been published in *Socialist Review*, *Z Magazine*, *Insurgent Sociologist*, and *New Studies on the Left*.

MARIANNE LAVELLE is a staff writer for the *National Law Journal* in Washington, D.C., who writes on environment, labor, and national science policy. She led the team for the 1992 investigation of the breakdown in environmental protection in minority communities.

ERIC MANN is the director of the Labor/Community Strategy Center in Los Angeles. He is the author of *Taking On General Motors: A Case Study of the UAW Campaign to Keep GM Van Nuys Open* (1987) and *L.A.'s Lethal Air: New Strategies for Policy, Organizing, and Action* (1991).

MARY MELLOR is principle lecturer in sociology at the University of Northumbria at Newcastle on Tyne, England. She is a member of the coordinating committee of the Red-Green Network and author of *Breaking the Boundaries: Towards A Feminist Green Socialism* (1992) and numerous articles on feminist eco-socialism.

VERNICE D. MILLER is a social-justice activist and cofounder of West Harlem Environmental Action. She helped produce the 1987 report *Toxic Waste and Race* for the United Church of Christ Commission for Racial Justice and was a member of the Advisory Board for the First National People of Color Environmental Leadership Summit in 1991. In 1992-93 she was a Columbia University Revson Fellow in Environmental Policy.

RICHARD MOORE is the coordinator of the SouthWest Network for Environmental and Economic Justice in Albuquerque, an organization of over thirty regional grass-roots groups working to empower grass-roots organizations to address the poisoning of their communities and to build a multiracial movement for environmental and economic justice. He has been a leader in the emerging environmental justice movement and a planner of the First National People of Color Leadership Summit on the Environment.

CHARLES NOBLE is a professor of political science at California State University at Long Beach. He is the author of *Liberalism at Work: The Rise and Fall of OSHA* (1986) and has a long-standing interest in workplace and environmental reform. He has written extensively on occupational safety and health policy.

JAMES O'CONNOR is the editor-in-chief of *Capitalism, Nature, Socialism: A Journal of Socialist Ecology*, and is a professor of sociolgy and economics at the University of California at Santa Cruz. He is the author of *The Meaning of Crisis* (1987) and *Accumulation Crisis* (1986).

JOHN O'CONNOR is the founding executive and chairperson of the National Toxics Campaign and director of the Jobs and Environment Campaign in Boston. He is co-editor with Gary Cohen of *Fighting Toxics: A Manual for Protecting Your Family, Community and Workplace* (1990). His most recent book is *Clean Dreams* (forthcoming).

MARK RITCHIE is the director of the Institute for Agriculture and Trade Policy in Minneapolis, Minn. He was founder and national chair of the Fair Trade Campaign.

JONI SEAGER is a Canadian feminist geographer currently teaching at the University of Vermont. She is the author of *Women in the World: An International Atlas* (1986). *The State of the Earth: An Environmental Atlas* (1990), and most recently *Earth Follies: Coming to Feminist Terms with the Global Environmental Crisis* (1993).

ROBERT WEISSMAN is the editor of the *MultiNational Monitor in Washington, D.C.*

BEVERLY HENDRIX WRIGHT is an associate professor of sociology at Lake Forest University, Winston-Salem, N.C. An environmental sociologist, she has written extensively on occupational health, race, and equity.

Index